"Breathtakingly honest, funny, and surprising, *I Never Promised You a Goodie Bag* is the unforgettable story of how, after a traumatic encounter in her twenties, Jennifer Gilbert found her way back to happiness. I couldn't stop reading."

—Gretchen Rubin, author of *The Happiness Project*

"It would be easy to get caught up in the chatty details of all the near-disasters Gilbert encounters in her fast-paced life in the world of event planning but this memoir is about more than cake and flowers."

—*Library Journal*

"This is a beautiful and brave memoir, full of hard-won wisdom. I read it in one sitting. It speaks from the heart and to the heart, showing us how in the hands of a clearheaded, strong woman hell-bent on not just surviving but thriving, the worst imaginable events in life can lead to unexpected joy. I was rooting for Jennifer Gilbert every step of the way, even though I knew that the book I was holding in my hands was a testament to her triumph of spirit."

—Dani Shapiro, bestselling author of *Devotion*

"What distinguishes Gilbert's memoir from the inspirational survivor pack is her willingness to share the bumps along her road to recovery. The story doesn't go predictably from devastation to bliss; she makes mistakes, suffers loss, endures heartache, and punishes herself by dieting and over-exercising. . . . A likable memoir that shows we can choose to be more than the sum of those events that are beyond our control."

—*Kirkus Reviews*

I Never Promised You a Goodie Bag

A Memoir of a Life Through Events—
the Ones You Plan and the Ones You Don't

JENNIFER GILBERT

HARPER

NEW YORK • LONDON • TORONTO • SYDNEY

HARPER

HarperCollins books may be purchased for educational, business, or sales promotional use. For information please write: Special Markets Department, HarperCollins Publishers, 10 East 53rd Street, New York, NY 10022.

First Harper paperback published in 2013.

Library of Congress Cataloging-in-Publication Data has been applied for.

ISBN 978-0-06-207600-7

13 14 15 16 17 OV/RRD 10 9 8 7 6 5 4 3 2 1

IN MEMORY OF JULIE SISKIND

There are stars whose radiance is visible on earth though they have been extinct. There are people whose brilliance continues to light up the world though they are no longer among the living. These lights are particularly bright when the night is dark. They light the way for mankind.

Hannah Senesh

Contents

Author's Note

This is a work of nonfiction. I have changed the names and identifying details of some individuals, companies, and organizations in order to protect their privacy. For the same reason, some events in the book are composites. I'll leave it to you, the reader, to figure out which ex-boyfriends, brides, and others with reason to fear have been granted anonymity.

Prologue

Keep calm and carry on.

S o, we have a little situation."

I was holding the bride's hand, looking up into her big brown eyes. She was standing on a stool so that her wedding dress could extend down to the floor, awaiting the massive tulle petticoat that would inflate all that satin to full-on princess proportions.

The bride knew something was wrong, and I could feel her fingernails digging into my palm. I smiled. "The first thing you need to know is that I will fix everything," I said to her.

The bride nodded. I had explained this to her before, just like I did with all my brides: things went wrong with weddings sometimes, and when they did, it was my job as the event planner to make everything right again. I always feel like I should wear a T-shirt that reads, "Chief Damage Control Officer."

I continued. "The second thing you need to know is that your tulle petticoat got caught in the trunk of the groomsmen's limo." The bride kept nodding, and then I added the kicker: "The limo drove away."

The bride was nodding more fiercely now, and there was a strangled sound in the back of her throat. The truth was, the last time I'd seen the bride's tulle, I was running down the streets of Miami, screaming at the limo driver while I watched the tulle being dragged and torn like the last sad piece of tissue in the bottom of my purse. But I didn't think painting that particular picture for the bride would be very productive at the moment.

Now the bridesmaids were getting in on the action. There was a collective gasp in the room, and one of them shrieked. The mother of the groom gripped my arm. Mrs. Lopez had survived Castro, but she looked like this might kill her. "Where is it? Are they bringing it back?"

Once again I recalled that tulle, dragged and ripped beyond repair through the streets of Miami. "Oh, honey." I sighed. "The tulle's not coming back."

Looking back on that day when I'd frantically chased down the doomed petticoat, I had an epiphany: while I was fixing things for other people, I didn't have to think twice about myself. Obsessing over every tiny detail of other people's most important events was what I did best. It was the perfect way to avoid thinking about the dark, scary void inside me.

I'd done a pretty good job of paving over that void with numbness, but every once in a while, on one of my bad days, the sorrow crept in, and I'd have to run and run to get away from it. So I welcomed other people's situations—their crises small and large. In fact no problem was too small for me to throw myself into fixing it. Like when the bride at another wedding insisted that we individually wrap hundreds of Sweet'N Low packets in their own little white envelopes because she didn't want to see any pink on her all-white tables. Or when we spent $50,000 on centerpieces for a corporate event—tall, gorgeous tree branches—only to find out an hour before the event that our speaker was less than five feet tall, and the audience would never be able to see him over all those branches. So we cut down every one of them using whatever knives we could swipe from the kitchen. Or when five hundred more people showed up for an event than the client had calculated, so I became a coat-check girl for the night. I did whatever it took to make a perfect event.

I was a master of the small stuff, but I really shone in the face of true disaster. That's when I got calm. When the party boat crashed into the dock, or the venue burned down two days before the event, I became the definition of grace under pressure. And when the bride's wedding dress was hanging a little (a lot) low, because her underskirt was currently taking a ride down South Dixie Highway? Whatever. It was all in a day's work.

When I woke up the morning of the tulle bride's wedding, it was one of my bad days. I could feel it even before my eyes opened—it was a familiar sense of dread, a physical ache of anxiety and fear coupled with a suffocating heaviness that filled the spot where my soul used to be. On a day like that, a runaway petticoat was just the jolt I needed. As I was skittering down the streets of Miami after the limo, I was the closest to happy that I could feel at that point in my life. I was in high-gear emergency mode, and all I thought about—all I cared about—was fixing this problem so beautifully that no one would ever know the wedding was anything but perfect. And a brilliant side benefit was that if I could hold this event together, then I could also hold myself together, at least for another day.

It was one of the first events that I had handled completely on my own, back when I was working at a small firm that specialized in wedding planning. The bride was a friend of a friend, and I'd become close to her. She was a little younger than I was, barely twenty-two. Her mother had died of cancer when the bride was just a child, and her father was now very ill. As a wedding planner you get used to acting in the role of family counselor without even thinking about it, and I could tell how unmoored the bride felt, as if she were already an orphan. A redheaded WASP from a non-religious Protestant family, she'd fallen in love with the son of a devoutly Roman Catholic Cuban-American family, and she was embracing her new life with open arms. This was her fresh start, her new family—and they were all there to watch her walk down the aisle of the biggest Catholic church in downtown Miami. The bride wanted everything about the wedding to be perfect, as if it might be a sign or a promise that her life would be perfect forever after as well.

Meanwhile, the dress. I remember walking into the room in the church where the bride was getting ready before the wedding. She'd already had her makeup and hair done, all curled in crimson ringlets. The Vera Wang dress she'd chosen had such full skirts that we lowered it over her head like an art installation while she perched on the stool. Then we did up the hundreds of tiny buttons that ran down the back of her dress. That was when we looked around for the petticoat. In the seconds during which eight pairs of eyes scanned the room for a pile of tulle, I could sense a rising communal panic. Murmurs started to bubble up. (*Whose job was it to look after the petticoat? What do you mean she left it in the trunk of the groom's car?*) Then there were louder hisses and the start of accusations, the kind that ruin relationships for years after.

The bride's older sister and I ran through the church out to the front, where the limos had been parked. Gone. Then we both caught sight of the retreating limo with the godforsaken tulle. I shook off my heels and did a barefoot sprint after the limo, screaming and yelling and waving my hands, but it was no use. The driver didn't see me, the limo was gone, and the tulle was in shreds—unusable even if we'd been able to get it back.

I hobbled back into the church with the older sister, who was already sweating in the Miami heat, hair sticking to her forehead and foundation starting to run. I gave her strict instructions: after I broke the news, it was her job to calm down the bride and keep her from crying all her makeup off.

My brain had already started to work, the wheels clicking and spinning as they always did when disaster struck. The tulle wasn't coming back, I knew that. But the bride couldn't wear the dress without a petticoat. Ergo, I needed to construct a petticoat.

I'd been (briefly) a fashion merchandising major in college, and I'd taken one sewing class—who knew it would come in handy? Still, I don't remember taking any lessons on how to make a tulle petticoat that would hold up a $5,000 dress for a long walk down an aisle. Luckily it was a

Saturday and stores were open, and we were in downtown Miami, where you can find pretty much anything. By that point I was sweating, blistered, and feeling a bit nauseated. Miraculously, within minutes I found a fabric store where I bought yards and yards of tulle, sewing needles, and white thread.

When I got back to the church, the bride was still standing on the stool. She hadn't dared step off and wrinkle the skirt, and there wasn't time to undo the hundreds of buttons and then do them up again. Now slick with perspiration, I got under the bride's dress. Swallowed under billows of satin skirt, eye-to-lace with the bride's Brazilian, I hand-stitched every inch of tulle to her garters, crinkling and bunching it up as I went so it would be dense enough to hold up the dress. It wasn't the most professional job in the world, but I knew it would work, and she could walk down the aisle with her chin up, and her dress floating around her exactly the way it was supposed to. Luckily, she also had an enormous veil that could camouflage any of the bumpier bits.

Vera Wang might have been horrified at the havoc I'd wrought on her design, but when I watched the bride glide through the church to her groom, a smile on her face and her eyes glistening with happiness—that's all that mattered. She was beautiful, and she was looking to the future with joy and optimism.

None of those things was true for me, and at that point in my life I believed they never could be. So instead of chasing the impossible, my mission became to surround myself with other people's joy, happiness, and hopes—that would have to be enough to sustain me. My clients—dreamy brides, loving anniversary couples, proud bar mitzvah parents, and demanding titans of industry—all helped me to carefully conceal the gaping hole inside. It's not supposed to be good for you to hide your fear and sadness. You're supposed to confront your problems. But I wouldn't have known where to begin, and I was terrified of going back to that dark place—the moment when I lost my joy. So for years and years, I survived event to event by burying my secret six feet under and dancing on its grave. For me, there was no other way.

Back then I had no conscious plan—it's only now, looking back, that I can see the wisdom in the road I took. As broken and shattered as I was on the inside, I polished up my outside and became an event planner. I surrounded myself with other people's beauty, happiness, and gratitude for life's milestones. Then something happened to me while I was planning all those happy experiences for other people. It started as a glimmer of longing inside me. Eventually it grew to a tingling, like pins and needles in sleepy limbs. I wouldn't call it hope, because that would have meant that I believed on some level that happiness could actually happen for me. No, it wasn't hope. The hope came later. The happiness came after that.

PART I
The Jen Suit

The greatest discovery of any generation is that a human being can alter his life by altering his attitude.

—WILLIAM JAMES

CHAPTER ONE

At Least

My childhood living room was an exotic landscape of bizarre and beautiful treasures from around the world—regal Chinese chairs, antique vases, hand-embroidered screens, and stunning cloisonné bowls. The walls were decorated with my mother's collection of framed antique kimonos, and the centerpiece of the room was a real tiger-skin rug with fantastically sharp teeth. My younger sister Rachel and I would dare each other to stick our hands inside the tiger's gaping mouth, and I still remember the breathless thrill I felt, as if absolutely anything could happen when our soft little fingers were in there.

My parents were both adventurers, and they passed the gene along to me. My father was in the import business back when you needed to actually go to the Far East to make your deals. Rather than leave his wife behind, as many husbands of that generation might have done, my dad always wanted my mother with him. And she didn't just shop or sit on a beach while my dad was in his buying meetings. She learned to speak Mandarin fluently so that she could translate for him and talk to the Chinese businessmen's wives.

My mother grew up in Brooklyn, sharing a bedroom with her older sister and younger brother, children of struggling shop owners who both worked in their store from 6:00 a.m. till 9:00 p.m. The only Jewish family in the neighborhood, my mother remembers kids throwing pennies at her when she walked to school. And there was no point complaining or vying for her parents' attention, because it was never effective. Her parents had enough else to worry about, so she stopped trying. The running joke about my grandmother is that my mother could have walked into the kitchen bleeding, with several fingers missing, and her mother would have said, "Let me make you a sandwich." From a very young age my mother knew she was on her own, and she wanted out. Even though her education was the least of her parents' concerns, she graduated valedictorian from high school, cum laude from Brooklyn College, and then paid for her own teaching degree. My mother then shocked her family by breaking off her longtime engagement to a college professor to travel around the world with a girlfriend, teaching English to diplomats' kids. The only member of her family even to have a passport, she wandered around Asia and lived on a boat around the corner from the Blue Mosque in Istanbul.

When she met my father, he'd recently taken over the family's import business and had broken off an engagement of his own with a girl from South Africa whose family wouldn't let her leave home. He saw my mother—beautiful, with a rocking body and long hair down her back—across the room at a singles weekend in New Hampshire. Six foot four and handsome (we call him the Jewish John Wayne), my father tried to impress my mother by telling her he'd just gotten back from Burma. She said, "Oh, yeah, I just got back from Turkey." He knew in that moment that he'd found his future wife. She was the adventurous spirit he'd been looking for, and he was offering my mother, a poor kid from Brooklyn, the riches of the world. They were engaged three months later, and married two months after that. There were two reasons things moved along so quickly: one, there was a cancellation at the banquet hall, so, as they've said ever since, "We got steak for the price of chicken" (words to live by). And two, my dad was scheduled to go on a buying trip, and he was determined to take my mother with him.

My parents' meeting and marriage set the tone for my whole childhood. We were a family that loved our travels. My dad was the only Jewish kid in Washington Heights who swam and sailed—he taught himself how in Boy Scout camp—and on family vacations he chartered a boat for two weeks at a time in the Bahamas or the British Virgin Islands. We'd go with another family with daughters—no captain or crew—and sail around the Out Islands like castaways. The adults would toss an inner tube off the back of the boat, and we'd take turns riding. I learned to snorkel before most kids learn to swim, and by age six I'd dive for sea urchins and knew how to wear gloves to peel away the outside with a knife. By the end of a trip I'd have decorated the entire boat with sea-urchin shells and silver dollars. Eventually my dad bought his own boat, and we spent three weeks at the end of every summer living onboard off Nantucket.

In the winter, it was skiing every weekend in Vermont (a four-hour drive each way from our home in Westchester)—until I got to be a teenager and started to rebel against leaving my friends. My parents felt incredibly strongly about those weekends and our vacations together. It was sacred family time, and to this day some of my most precious memories are of scampering around islands and racing down slopes with my sister Rachel. Those were the feast times.

The famine times were when my parents went away for my father's selling trips. For three weeks at a time, several times a year, they were gone, and we kids—Rachel, our youngest sister Marissa, who is ten years younger than I am, and me—were left in the care of a nanny or housekeeper. Every time they left us, my mother agonized. A part of her wanted to stay, but a part of her wanted to go, too. This was thirty or more years ago, and in our privileged suburb the moms stayed home. They certainly didn't leave their children while they went abroad for long periods. But my father insisted that he needed her with him, and she saw it as her job and her responsibility to go. She also loved the experiences. If she wasn't traveling with my dad, then she was studying some foreign language with a tutor at our kitchen table.

We knew the trips were required of my father for his work, and it instilled in me a strong work ethic from a very young age. Growing up, I always had a job or was hatching some kind of entrepreneurial plan. Once it was flooding the backyard to make an ice-skating rink (for which I'd collect admission, of course). Another time it was putting on a circus in our backyard. I was the star performer—the only performer—and my middle sister Rachel was my trusty sidekick and ticket collector. In high school I started a shockingly lucrative business selling necklaces from a huge overage of beads that my dad had imported. He was thrilled. It was also mandatory that I waitress one summer. My dad felt very strongly that serving hungry people was an essential lesson if you wanted to be a success at anything in business.

Even though I valued my dad's work, my parents' trips away from us felt like little eternities, and it only got worse as I got older. When I was going through all my adolescent changes and I needed my mother most, it didn't feel like she was always there. And because I was so good at taking care of myself, I'm sure it didn't even occur to them to check up on me. I was smart, and I brought home good grades. I played field hockey and lacrosse. I skied, I sailed. I was even an expert at riflery. I was a strong, independent little girl, and I grew into a self-sufficient young adult.

But like any kid, I made my mistakes, and my parents weren't necessarily there to pick up the pieces. When I ended up in the ER after an accident at age twelve, it wasn't a parent who came to get me; it was the family friend whose name was listed under my emergency contacts. I had fallen at school while running on concrete and torn up my hands and hips. Actually, I'd fainted before falling, because for days, all I'd eaten were gummy bears, popcorn, and Cheerios—and then I'd biked several miles to school. That was my adolescent idea of a healthy lifestyle, and it seemed like a really good idea at the time. Lying in that hospital bed not having my parents around was the first time I felt something close to homesick—only my parents were the ones who were away from home.

My friends' parents seemed to hold their daughters' hands through every milestone. I went into the SATs completely cold. My parents had registered me for the prep course, but they never checked that I was actually attending, and so I just didn't go. They were on a trip when I took the test, and the night before, my sister Rachel threw a loud party. The next morning, when other kids' parents were making a special breakfast for them and driving them to the test, I set my own alarm and drove myself. When it came time to apply to colleges, I spent hours at my friend Karin's house writing my essays and filling out the applications, and then she proofed my essays for me. My friends and their families were the support group I created to get me through the times I longed for my parents and a consistent childhood. The moms made me dinner, always including me in their rituals and standing in when my parents were absent.

Karin was one of my five closest friends in high school, and the others of our core group were Dianne, Julie, Carolyn, and Jenny. Everyone called us the six-pack, because we were inseparable. The night after we all took the SATs, I decided that we should throw the high school party to end all high school parties. I'd been to parties at other kids' houses, and I'd taken careful note of all the pitfalls—I knew what got the cops called, and even more important, what got the parents called. I'd been to crazy parties where kids decimated the liquor cabinets and desecrated the bedrooms. The family's home would be such a wreck afterward that no amount of frantic cleaning up could ever cover the tracks.

Not at my house. I perfected the art of throwing a chaperone-free party without getting caught. I'd even learned that the way parents figured out their kids had thrown a party was by counting the rolls of paper towels afterward. If too many rolls were missing, that meant a lot of drinks had been spilled over the weekend. So I made sure to count the rolls and replace exactly the number that we'd used. (To this day, my mother finds this hilarious; she says that even then I ran a tight ship.) I got the six-pack together, and I divided up the jobs. One girl had an older brother who hooked us up with kegs of beer. I was dating the star football player, so he recruited his teammates to be my bouncers (pity the peewee freshman shaking in his boots when one of those football players asked for his cover charge).

Meanwhile, I worked on the invites. I prided myself on making friends with everyone—popular girls, jocks, nerds, stoners, artists. So I got the word about the party out to all my diverse circles of friends and told them to invite their other friends as well. The message "Party at Jen's house" eventually got passed to (easily) six hundred teenagers in a twenty-mile radius of my town.

The entire party was held outside in my backyard. It was a cold fall evening, but there was no way I was letting those kids into my house. When people needed to use the bathroom, they waited on line and then one of my girls escorted them to and from the bathroom door. Every other door in the house was closed and locked. I stayed outside—dressed in my cargo pants with all the pockets stuffed with money. I collected $10 from the freshmen, less from the upperclassmen, and if I really liked you, then you got in for free. We made thousands of dollars that night. I paid for my prom ticket, the dress, and the night out after.

I wasn't a big drinker anyway (and I hated beer), but I never drank at my own parties. So when the cops finally did come, I was sober and managed to conceal my terror when I answered the door. I remember the incredibly nice officer standing there with his partner, both of them kind of laughing. He was about my dad's age, and he said, "Look, I know my son's back there somewhere, and I don't want to go in and embarrass him. So just turn the music down, and everybody goes home in a half hour."

My girlfriends all slept over—each of them making the excuse to their parents that they were staying at someone else's house. Then the next morning, my friends and I and my youngest sister Marissa (who was a tender six or seven years old at the time) each took a five-foot-square area of our backyard and picked up every bottle cap, cigarette butt, and scrap of litter. That night turned out to be just the first of many "parties at Jen's house" when Mr. and Mrs. Gilbert were out of town. Now of course my parents credit themselves with giving me my start in the event-planning business.

They're not entirely wrong. And I know I have them to thank for my taste for adventure, too. As a kid I had already figured out how to be resourceful and independent, and capable of going through my life without having to ask for help all the time. When most seniors were terrified of leaving the nest, I was unfazed by my impending freedom. One of the main reasons I chose to go to the University of Vermont was because I loved skiing so much, and I'd spent such an important portion of my childhood racing down those mountains. I loved sailing just as much, so my senior year in college I was accepted into the Semester at Sea program and traveled around the world on a boat.

Throughout my college years I was always meeting new people, and my love of organizing events stayed with me. I was president of my pledge class for my sorority, Kappa Alpha Theta, and then I was voted "service chair," which meant that I was responsible for planning our large parties for charity. I had a wide circle of friends—girls on my dorm hall, in my sorority, and on the women's lacrosse team that I tried out for my freshman year. But I needed a new support system—my college family. My best friend at that time was Nicolette, and we became very close with another girl on our hall named Andrea. Later the three of us became roommates, both of them a year ahead of me. I was the sloppy one (and the procrastinator), and Nicolette and Andrea would close the door to my room because it was such a disaster area. The three of us were attached at the hip, and my most vivid memories of college are of Nicolette, Andrea, and me, out until the wee hours, coming home buzzed, and knocking on the door of the campus bakery that was located (conveniently enough) on the ground floor of our dorm. They'd open up for us, and we'd scarf down free cookies and brownies.

After they graduated, Andrea went to work in New York City, and Nicolette moved to London. As my own graduation approached, I knew that Dianne, Julie, and Karin from my high school six-pack were also getting temporary work visas and moving to London. There was no way I was going to let them have that adventure without me. Nicolette found me an apartment and a roommate close to her flat, while my other friends were just a few tube stops away. My parents had no problem with me going so long as I knew that they weren't bankrolling the trip, so I made my own money while I was there and saved up as much as possible. Then when my work visa was about to expire, I traveled all over Europe with my friends—backpacking, hitchhiking, flying by the seat of my pants. I was the girl who led the way and took charge, deciding on the spur of the moment to head west to Portugal or south to Spain, and everyone else followed along.

Eventually my savings ran out, my visa expired, and it was time to go home and figure out what I wanted to do with my life. I wasn't sure what exactly that was, but it would look something like this: I'd be in business or finance, taking corporate America by storm. I'd live in Manhattan and wear a Donna Karan onesie under my power suit. It would be a fabulous life.

I'd been back from London just a week when I took a train into downtown Manhattan to visit Andrea. While I was away, Andrea and I had talked once a week so we wouldn't miss any part of each other's lives. We couldn't wait to see each other.

I was a smiley twenty-two-year-old with long straight hair and dimples. I remember exactly what I was wearing—black flats, a tan linen wraparound skirt from Ann Taylor, and a black T-shirt. As I walked along, I caught my reflection in store windows, checking out how I looked. My reflection showed a sophisticated woman, and I felt proud and invincible after my year working abroad.

I was always frugal, and even though I'd never ridden the subway in Manhattan before, it didn't even occur to me to take a cab. I figured I'd been taking the tube in London for the last year, and how hard could it be? I did pretty well, but I missed my stop and got off at Canal instead of Houston, where Andrea lived. So I walked up Varick a few blocks, then made a right on Houston on the wrong side of the street. I was looking at the building numbers, realized my mistake, crossed the street again and then doubled back to Andrea's apartment. The doors to her building were propped open because a couple of men were there working on their motorcycles. I cruised in the front door without buzzing to get in, and smiled and waved at the men. I made a right turn down a long hallway to Andrea's ground-floor apartment. There were other apartments, but the hallway was empty.

I rang the doorbell, and then I heard a noise. I looked to my left, and walking toward me down the hallway was a man—about 250 pounds and carrying a backpack. Red shirt, black pants. It was odd, I thought, the way he was walking right up against the wall, on a beeline to run into me. I rang Andrea's bell again. The man's head was down, looking toward the floor, and then his gaze—just the eyes—shifted up, and he was looking straight at me, through his eyebrows. That's something I will never forget: those eyebrows and that hairline. What I saw in those eyes was terrifying. It wasn't lust, or drug-fueled rage, or insanity of any kind. It was pure hate. A shudder rolled across my body, immediate and involuntary. Now I was frantically ringing the doorbell. *Hurry up, hurry up.*

Thinking back on it now, the most amazing thing about what happened next wasn't the attack, it was the way my brain worked. I kept *thinking* the entire time. Never once did I stop calculating what might come next. Never once was it a blur.

The first blow landed on the side of my head, then came two or three more fast, pounding blows. Bam. Bam, bam, bam. I instantly threw down my bag, thinking he wanted my purse, and I thought, *Take it, just for God's sake stop hitting me.* That was when the real terror struck me, because he didn't stop. He didn't want my purse—he wanted something else. The next blow didn't feel like the others, which I'd thought were punches—this one landed on my thigh, and then it penetrated and sank into my flesh. That's when I realized: he hadn't been punching me this entire time, he had been *stabbing* me. I was on the floor now, curled up and trying to block him from stabbing my face. That's why I still have scars on my hands and between my fingers. Later, I would find out that he'd used a screwdriver.

Then the door opened. I looked up and saw Andrea's face, and I thought, Oh, God, thank God. She saw me lying on the floor covered in blood, straddled by a man whose arms were pumping up and down while he stabbed me over, and over, and over again with the screwdriver. By this point my ear was ripped, my neck was bleeding, and I had stab wounds on my head, hands, and legs. In her panic and horror, Andrea closed the door.

Now I was alone again, in the hall, with the man who was about to kill me. I remember yelling, "Oh my God, Andrea, help me, he's stabbing me." I fought. I kicked, I screamed, and somehow I got him off me, and I got to my feet. My skirt was off then, and I thought, That's it—*he wants to rape me.* In that moment, Andrea opened the door again.

I leaped for the opening and ran down the hall of her apartment—to what I'd hoped was safety. But it wasn't. I turned to look behind me, and the attacker had pushed past Andrea, flailed at her with the bloody screwdriver, and was running after me down the hall. Andrea, who said later that she had feared for her life, dashed into her own bedroom and locked the door behind her.

As I ran away from the attacker, I remember seeing a marble chess set, and I was so desperate for something to fight him off with that I picked up the chess pieces and threw them behind me. Once I was at the end of the long hallway, I was trapped in Andrea's small, square living room with nowhere else to run.

I wedged myself on the sofa and shot my leg up to kick the attacker in the balls. Instead, he grabbed my foot, spun me around so that I was flat on my back, and straddled my stomach, facing away from me. Until that point, I had continually been thinking, What does he want? When he ignored my purse, I knew he didn't want to rob me. Now that I was pinned underneath him, facing his massive back while he tried to pry my legs open to *stab in between them*, I knew that he didn't want to rape me, either. This man didn't want anything from me—he wanted to kill me in the most violent way possible. I thought: I am alone, and I am dead. I clenched my legs so tightly closed that I would have trouble walking for at least a week after. I screamed and I beat and pounded on his back.

I felt no physical pain in that moment, even though I probably had more than thirty stab wounds. I have scars all over my body from the attack—scars between my fingers, on my neck and throat, all over my legs—but I have no memory of feeling pain. Andrea's white sofa was red with my blood, and still I couldn't feel anything physical. What I felt was an instinctual fierce, bottomless urge to fight, no matter how convinced I was that I was dead.

Then, suddenly, the attacker was off me. I heard him run down the hall, and the door to the apartment slammed behind him. I had no skirt on, and there was so much blood dripping from my body that my feet slithered and slipped in my shoes. I wasn't crying, though. I walked down the hall, past Andrea's locked bedroom door, and then I turned the locks on the front door.

After I locked the door, I screamed, *"Andrea, Andrea, open the door!"* and she said, "Is he gone?" I said, "Yes, he's gone, open the door." When she opened her bedroom door and saw me, she started shrieking. I was standing there, underwear soaked with blood, stab wounds all over my body, and one ear hanging. I must have been in total shock, but I calmly said, "Call the police, call an ambulance, call my parents." I found out later that she told them that I'd had an accident and suffered a few scratches, but I was okay.

After Andrea made her calls, the disbelief hit me. I said to her over and over, "Did this just happen to me? What just happened to me?" Then I had to lie down, right there on the floor of the hallway. I asked her, "Did he get my face?" She didn't answer, and I said again, "Andrea, did he get my face?" She said she didn't think so. Then I said, "Did he get in between my legs?" She said, "Gilbs, I don't know . . . there's so much blood."

Years later, it would be a shocking revelation to me that Andrea's four roommates were actually inside the apartment when I was attacked. When Andrea closed the door on me, she'd tried to tell them what was happening, but she couldn't speak. She hadn't even been able to get the words out while I fought for my life in the hall. After I ran inside, chased by the attacker (and Andrea had locked herself in her own bedroom), instead of together helping me to fight him off, the four roommates had run to get help from some male family members who lived in the building. As I lay on the floor of the hallway, minutes after the attacker had already run out, two of those men came in, holding baseball bats. Then they took off after the attacker and chased him to the subway, but lost sight of him when he jumped onto the tracks.

Before running away, my attacker had the presence of mind to tell the two men working on their motorcycles out front that a woman was being attacked inside. The two men said to the police that when the attacker followed me inside, they'd assumed he was with me—that's how close to me he was. Apparently he'd followed me all the way from the subway. All that time I'd walked from the wrong subway stop, crossed the street, and doubled back, trying to find my way, there he'd been—right behind me.

The ambulance came, and I was strapped to a gurney. As I lay in the ambulance, I remember the face of the paramedic looking down at me. I thought: I'm going to die. I was one hundred percent certain. I said, "Tell my family I love them." I told myself that I had lived a good life. The paramedic's face was just inches from mine, and he said to me over and over, "Hang on, Jennifer, hang on, honey." I was fighting unconsciousness.

I was pushed through the doors of St. Vincent's Hospital, and five doctors surrounded me, running alongside the gurney as I was rushed through the halls. I saw the lights overhead and their faces looking down at me and I heard a voice saying, "Multiple stab wounds. Stat."

I was alone, in deep shock, and convinced that I was dying, but somehow I knew that I had to stay focused and awake long enough to tell the doctors what had happened to me. They were afraid to undress me for fear that the attacker had gotten my internal organs and they'd open up a gushing wound on my torso. The doctor said, "Can you tell me where the wounds are?" I said, "Everywhere. My face, my back, my legs, my stomach." When I'd finished telling him all the places I'd been stabbed, he said, "Okay, honey, we're going to take care of you." I looked at him and said, "I'm gonna go now." And I was out.

. . . .

This is what people said to me after the attack:

At least he didn't get your face.

At least you're alive.

At least you weren't raped.

I learned that this is what "at least" means: Move on. Get over it. Let's not talk about it. It could be worse, so it must be better.

This is what I asked myself after the attack:

Why did I miss my stop?

Why did he follow me?

Was it something about me, how I looked, how I was dressed?

Did I smile at him or say hello?

What if I had turned around and seen him, gone into a store, told some big man on the street that someone was following me?

What if I had seen him in that window when I checked out my outfit?

Why didn't I notice I was being followed?

Why didn't I notice he was right behind me?

Why didn't I notice?

What terrible thing have I done in my life that I deserved this—have I not been a good enough person?

In what universe is it okay for this to happen to someone?

The questions ran around and around in my head, a constant series of mental laps. Did I do something? Did I not do something?

There was a sharp, medicinal smell, and I opened my eyes to bright lights overhead. I was awake, which meant I was alive. Oh my God, I was alive. The doctor saw that I was conscious, and the first words I could think to say to him were, "I'm here."

I was silent and calm while they finished sewing me up. I don't know exactly how many stitches it took. Most of my wounds were punctures. I was told I was lucky that the attacker had used a screwdriver and not an ice pick or a knife—if he had, I'd be dead. As it was, the screwdriver sank into me, but didn't slice on its way out. But the force of those punching wounds was like being beaten as well as stabbed. If I had to guess, I'd say there were something close to forty stitches—at least fifteen of those to sew up my neck. They used surgical glue for the puncture wounds. For weeks after the attack, I'd need to use a walker.

I remember the first time I saw my parents after the attack. I was already out of the ER and in a hospital room, and I must have been in a tremendous amount of pain. The nurse had pulled a curtain around my bed, so I heard my parents' voices before I saw them. I'd been so alone, so convinced that I was going to die. I'd been ripped to shreds and some-what sewn back together. Andrea had told them that there were only a few scratches, and now they were about to see their daughter bleeding through the hospital sheets. They still didn't know that someone had tried to kill me. So the first words I said to them—the first words it occurred to me to say—were: "I'm fine, I'm fine, I'm fine."

CHAPTER TWO

Turn the Channel

This is the moment that I realized my life had become surreal: it was three days after I was attacked, and I was with my mother in my parents' living room. My puncture wounds would take months to fully heal, and they were still seeping and bleeding through the bandages, so I sat on a thick towel to protect the sofa. My mother and I were watching *General Hospital*, and when it ended, the familiar face of the local news anchor popped on the screen to do a quick promo for the evening's top story.

I have very few concrete memories of those early days after the attack, but I remember that promo: "Screwdriver-wielding maniac loose in New York, details at five." Later, when the news came on, I watched a reporter interview people who were in the apartment building when I was attacked. One remembered how I'd screamed, "Oh my God, he's stabbing me, call the police."

Watching that news report, I froze. I peeked at my mother without turning my head, surreptitiously gauging what my own response should be to seeing this on television. Then she shuddered, picked up the TV remote, and turned the channel *without saying a word*. Her face was unreadable and focused on the screen. The whole scene was surreal. I didn't blame her for turning the channel, but I remember thinking, Did all of this really happen to me? I truly thought I was losing my mind.

Every second while I was being attacked I had been doing something—kicking, yelling, calculating my next move. But watching that newscast, I was speechless, unable to do or say anything. I just stared in paralyzed disbelief at what my life had become. I had become a top story—the headline that made people gasp and turn up the volume on the TV a little louder, or shut it off altogether. The irony was that while people all over New York were talking about me, my family was living in silence.

My attack—and the attacks of two other women by the same man—spurred a regional manhunt. Guardian Angels handed out composite pictures on street corners all around Manhattan. These days people are inured to news of violent attacks splashed across newspapers, TV, and the Internet. But in 1991 my attack was shocking—and massive—news. It was the headline of every paper, and there were daily recaps of the search for the attacker on the evening news. Although my name was withheld, I still felt deeply exposed.

One day not long after the attack the telephone rang, and I answered it to hear a strange voice on the other end. The caller's first words were, "I'm looking for Jennifer Gilbert." I stayed silent, in shock. He spoke again. "I'm looking for Jennifer Gilbert, the victim of the subway attack." He said that he was calling from a newspaper. My pulse raced, but I calmly said, "There's no Jennifer Gilbert here. You have the wrong number." It was only after I hung up the phone that I started to hyperventilate.

After that, news vans began camping outside our home. The only way for me to get out of the house without being seen was if one of my family members pulled the car right up to our front door. Then I'd have to dive in and hide in the backseat under a blanket.

That kind of publicity around a personal tragedy would have been unwelcome under any circumstances, but in my case it only added to my feeling of being under assault. I lived in constant terror during that time. Irrational or not, I thought that the attacker might come and find me to finish the job. I had thrown aside my purse during the attack, and I was gripped with an unshakable paranoia that he might have seen my identification—that he might know who I was and where I lived. If the reporters had found me, why not him?

I was a physical wreck as well. I had bandages on my face, and I could barely open my jaw because I had been hit so many times. In addition to the puncture wounds all over my butt and legs, there were more on my hands, neck, and one side of my head. My muscles were so traumatized from clenching my legs together while the attacker had tried to pry them apart that I couldn't walk unassisted. I was paralyzed inside and out. I couldn't make it up and down the steps to my own bedroom, so I slept on a pullout sofa in the den downstairs.

My physical discomfort would have been enough to keep me awake at night, but I dreaded the night for other reasons. The only way I could rest was if the lights were on, the door to the room was open, and my father was sleeping outside. Even then, I had horrible, screaming nightmares. I was always being chased, always fighting for my life. I was running, falling, running. There was a person chasing me, sometimes an outline, sometimes a face—the face of the attacker. I was afraid to go to sleep, afraid to dream. In the twenty years since the attack I've recovered in almost every other way, but I still don't sleep well.

I developed strange phobias after the attack, things I had never been afraid of before. They always involved situations where I didn't feel in control of my physical surroundings.

I was afraid of flying in airplanes. I love to travel, so despite my anxiety, I swallowed my terror and flew in a lot of planes.

I used to sweat and get heart palpitations when I drove through tunnels. On bridges over water I had to roll all the windows down, because I was afraid that if the bridge collapsed I wouldn't be able to get out and I'd drown. I live in New York, so I had to force myself to drive over a lot of bridges and through a lot of tunnels.

I had a fear of elevators. If I was alone on an elevator and someone else got on—someone who got my adrenaline pumping and the hair standing up on my arms—I had to get out at the very next floor.

For three years after the attack I lived with those phobias, and I hid my fear. Finally they all faded away.

Now that I have children, I've developed two new phobias. I have a terror of my kids drowning or choking. When that fear comes over me, it's as if my body remembers what it was like to be under attack. My ears close, and my rational mind competes for control with my body's panic response. I know already that these new phobias are never going away. So I make sure my children get swimming lessons, and I swear I'll cut up their grapes until they go off to college.

In the weeks following the attack my parents drove me wherever I needed to go, and indulged me by checking the trunk to make sure no one was hiding inside. My mother suffered through police visits, and she sat beside me while I described the attacker for the police artist. She refused to allow the police to take pictures of me after the attack. She wasn't about to let some stranger strip her daughter naked and take photographs—or, God forbid, risk having the photos released publicly, plastered all over television newscasts and newspaper accounts. Instead my mother sent my father out to buy a Polaroid camera, and then she carefully, painstakingly took the pictures herself to have on file for any legal proceedings should the attacker ever be caught. She made sure that they were all close-ups, revealing as little of me as possible. Other than the hospital staff, she was the only person who saw all of me after the attack. Even I don't remember what I looked like—but she does.

My parents were always physically nearby during those days, but neither of them was able to talk to me about what had happened. For my mother, the worst-case scenario—that I might have died—was too horrible to imagine. So her internal mantra became, "Thank God my daughter is alive, and let's not speak of this again." She wanted to lock the whole thing out. My father—a man so tender that my mother would tease him about it—never once let me see him cry.

Emotionally, I was still in a state of numb shock and disbelief, and not talking about what happened only made the whole thing seem that much more impossible ever to comprehend. I knew that I would never be the same again, that my life would forever be divided into a before and an after. But with no one around me acknowledging that—because why on earth would they want to?—I was left unable to grieve for the happy-go-lucky Jenny Gilbert that I'd been before the attack. I couldn't mourn for her, or admit (out loud, anyway) that even though my heart was beating, my soul was dead.

I don't remember looking in a mirror for at least eight months. I'm sure I must have, but I have no recollection of the face that looked back at me. I do remember that every time friends and family looked at me in the weeks after the attack, the expression on their faces was a mixture of pity, horror, and sadness at what had been done to me. Often they'd cry.

Their emotional responses always made me nervous—upset that I'd upset them—and I'd rush to tell them that I was fine, exactly the way I'd done when my parents first came to the hospital. When the visitor inevitably asked me what happened, I'd go into a rote recitation of the attack, almost like I'd been split into two different people. From a distance, the Jennifer sitting in my parents' living room described what had happened to that other Jennifer who'd been attacked. If things ever got too emotional, my parents would step in and change the subject. I never spoke about my night terrors, or about how I felt empty inside, or about my conviction that nothing good would ever happen to me again. I never spoke about how frightened I was that the man who attacked me was still out there somewhere.

I received hundreds of letters from family and friends, and although I was barely holding myself together, I (insanely) wrote a thank-you note for each one. Neighbors dropped off food, pies, and cakes. The cards and casseroles filled the house like condolences, and it felt as if I had died. People came to visit, perching on our living room furniture like they were sitting shivah for me, only I was right there. I formed sentences, I ate and drank, but I was just going through the motions, imitating the movements of a living, breathing human being.

Every terrible loss has its own shape and color, and I don't blame anyone for not knowing how to respond to mine. I knew the people who called or came to visit were trying to reach out to me, it's just that they had no idea how to do it or what I really needed. What we all want is to make people feel better when they're suffering, and we think the best approach is to encourage fortitude, and focus on the positive. Over and over I was told:

Thank God you're so strong, you'll be able to get through it.

God only gives you what you can handle.

Whatever doesn't kill you makes you stronger.

I'm sure this was meant to be encouraging. But the message I received was that I should feel lucky to be blessed with such resilience, and that they expected me to bounce back, just as good as new. Meanwhile, I couldn't imagine leaving my house without an armed escort.

Some friends wanted to reach out to me by sympathizing with what I'd been through. They'd tell me stories about terrible things that had happened to other people, in the hope that I wouldn't feel so alone in the terrible things that had happened to me. They'd say things like:

I know just how you feel, because I was mugged.

I know just how you feel, because my friend was robbed at gunpoint.

I know just how you feel, because my sister was in a car accident.

Even in my fog of disbelief, I recognized that no one actually knew what I had gone through. I had been picked—targeted, followed, and attacked with the express intention of murder. It hadn't been an accident. And while I was being almost-killed, I'd also been abandoned. And I hadn't only been battered physically; I'd had my entire worldview altered. I'd learned that the universe had nothing good in store for me. I'd been marked with a big V for *victim*, and if something so terrible could happen to me once, why couldn't it happen again?

I'd nod when people told me they knew how I felt, but they couldn't possibly know. We all have our own histories and our own unique ways of absorbing pain. No one else was inside my body, and no one could feel what I felt—either during the attack or while I was coping with its aftermath. In the end, all the failed attempts to reach out to me broke my heart. I wanted so badly to be reached, but none of it worked for me. There was no solace to be found in other people, I concluded—I had only myself. My unabating pain became my own little secret.

I felt physically exhausted by the emotional effort of trying to hold myself together for everyone around me. Of course there were people who urged me to go to a support group. But I didn't want to hear about other people's horrendous experiences. I didn't want to feel better about my situation because someone else had it worse. And I couldn't imagine how it would help me to know that there were lots of other victims out there who had terrible things happen to them—I didn't want to hear all the other ways in which the world was not a safe place for me.

I didn't come from a family that ever turned to therapy, but things got so bad that summer that I gave it a try. What I really needed was a crisis therapist—someone to get me through the immediate trauma of surviving a near-death attack. But instead I was given a referral to a family therapist. During our first session, he decided that my parents needed to be present and suggested we have a session with all four of us—him, my parents, and me. It was a huge mistake. I went in still trying to protect my parents from the pain I was feeling, and he was challenging that dynamic. It quickly deteriorated into guilt and recrimination, the last thing that any of us needed. I needed to preserve whatever relationship I did have with my parents at that moment. Exposing all the old hurt just made the new hurt more unbearable. No amount of blaming my parents was going to bring back my joy, or resuscitate that optimistic girl I'd been on May 30, before I walked into Andrea's apartment building. So that was the end of therapy for me.

. . . .

There were two people who gave me some sense of relief during that time, and I would never have expected it from either of them. The first was my sister Rachel, three years younger than me and just home from her junior year in college.

Rachel and I had been thick as thieves when we were little kids, right down to our matching bathing suits. I was ten and Rachel was seven when my youngest sister Marissa was born, and while we competed for our parents' attention when Marissa came along, we stayed attached at the hip. By the time I got to high school, though, we had drifted apart, both of us trying to find our own way through adolescence. I'm sure I became the classic bitch of an older sister. She followed me to the University of Vermont, but we had different groups of friends, and at most we'd wave to each other in passing. When Rachel came home from college after my attack, she had her own summer plans, and they certainly wouldn't have involved taking care of me. But she stepped in to take turns with my dad sleeping near me every night. Constantly shocked awake by my thrashing and screaming, she comforted me through my sweaty night terrors. During the day, she stayed with me in the house because I couldn't stand to be alone. If she had to run an errand when my parents were out, she took me along with her, never once complaining. She stepped in where my parents couldn't. She didn't even look at me like I was crazy when I flinched because someone walked too close to the car.

There was one other person who seemed to understand instinctively what I needed. Laura and I had been friendly in college, but we hadn't been exceptionally close. When I went to London we wrote letters back and forth, and then after the attack she came to visit me at home. In her body language, her facial expression, her words, everything she did and said, she was completely different from anyone else (no doubt this was at least partly thanks to the fact that she was getting her master's in social work). There was something about the way she approached me that instantly made me feel that it was okay not to be strong, that it wasn't my obligation to feel lucky to be alive, and that she wasn't there to compare what had happened to me to other traumatic events. She was just there for me, and she didn't bother with eloquence. After she heard what happened, I think her exact words to me were, "That just *sucks*." And it was as if for a brief moment I wasn't delusional. Yes, it did suck! It really, really sucked.

Months later, I'd often sleep over at Laura's apartment in the city. I was still terrified of sleep, so at bedtime she'd come down from her loft and she would pat me on my back. That was her goodnight to me, and her inoculation against the terror that nighttime always held for me—*pat, pat, pat*. I was still in survival mode in those years, ready to fight or flee, and I hated to be touched. Her *pat, pat, pat* was about as much as I wanted or could handle; it was my protective covering for the night. Laura became one of my best friends, and one of the very few people—numbered on fewer than the fingers of one hand—for whom I didn't have to pretend. It shocked me that a new friend could care for me in this way, when people I'd known for years seemed at a loss around me. I guess it was her inherent wisdom, because twenty years later, she would become Dr. Laura Berman, a leading expert on relationships and intimacy.

Nearly my entire group of close friends—including Nicolette and most of my high school six-pack—were still away in London and Europe. I would have turned to them more if they'd been home, but it was torturous to try to tell them my story on a transatlantic call. One of my very dearest friends, Deanna, didn't find out about the attack until months after, because she was traveling in Italy and I had no way of reaching her. Another close college friend, Amanda, was stateside and could have come to see me right away. After college and before I'd left for London, Amanda and I had spent one of the best summers of my life, sharing a house with a bunch of guys in Nantucket and supporting ourselves making sandwiches. I called her after the attack, but she just couldn't handle it. She never called me again, and never came to see me.

When Andrea visited me after I came home from the hospital, she burst into tears. She said she was sorry. I loved her, I told her that I didn't blame her, and that it wasn't her fault. I reassured her that none of us can know what we'd do in such a situation until we're in it.

Other friends and family who heard about the attack were horrified by Andrea's actions that day. They would ask me how she could have hidden behind a locked door while her friend was screaming for help on the other side. In Andrea's defense, I would always say the same thing that I'd told her: "None of us can know what we'd do in that situation."

Of course I would have hoped for someone to help me, especially one of my best friends, but I knew she couldn't help her actions in that moment; it's just how she reacted to panic. So I rushed to make Andrea feel better. It was the only conceivable choice for me. If I hated her, or blamed her, I'd lose her as a friend, and I couldn't allow that to happen—I had already lost enough.

The manhunt for the attacker dragged on throughout the summer. Meanwhile, I chewed the inside of my cheeks while I slept, and I'd wake up with the taste of blood in my mouth. I couldn't keep food down and threw up regularly. When we traveled back and forth into the city for the police interviews and composite pictures, I hid under my blanket in the back of the car, certain that the attacker could be anywhere, and I was a walking target. I was afraid of New York, of every stranger I encountered. I was afraid that I wouldn't know when I was in danger. Potential threat surrounded me—it hid, it lurked, and I didn't feel safe anywhere.

For the most part my father drove me into the city, because my mother felt oddly antagonistic toward the police, as if they were prolonging our agony. Finally, toward the end of the summer the police called to say that they'd picked up a suspect, and they wanted me to come in for a lineup. We were all brought in—Andrea, the other women who were in the apartment, and the two men who'd been outside fixing their motorcycles—although they kept us separate.

When I was taken in for the lineup, I saw that there was a one-way mirror looking out on the room where the suspect would be, and no amount of rational reassurance could convince me that the attacker couldn't see straight through that glass. I hugged the back wall, I was so terrified, as if every panic response I hadn't felt during the attack was seizing me now. Just the knowledge that I was in the same building with the person who could have been HIM—much less on the other side of a thin layer of glass—petrified me.

They brought the men in, and each of them had a large bandage completely covering his eyebrows, and a hat on to conceal his hair. I had been told the attacker had tried to disguise himself by shaving off his eyebrows and hair completely.

I knew the attacker's face—it was in my mind—and my memory of his hairline and eyebrows had been strong. I'd described him to a T to the police artist whose composite drawing had appeared in newspapers and posters around the city. But there standing right in front of me, months later, were eight men. They were all the same height, weight, and complexion, with no eyebrows and no hair to see. How could I know for sure? In the room with me were policemen, detectives, probably a few lawyers, but as far as I was concerned it was just me and those men on the other side of the glass. I was shaking from the pressure. I scanned the row, and two faces immediately jumped out at me: I thought, Number four, or number seven. No, number four . . . That was the man who'd tried to kill me; at least I was pretty certain.

These are the thoughts that went through my head in that moment:
What if he's staring at me when I put him away?
What if he knows who I am?
What if that's not him?
What if he's still out there?
Why are they making me do this?

I believed the attacker was number four, but I debated. It's a wonder I didn't throw up on the nearest officer and run from the room. I told the officers that the attacker was number four or number seven. A detective asked me, "If you had to pick just one, which number?" Then—overwhelmed, claustrophobic, desperate to get out of that room—I picked number seven. Right afterward I told the detective that I thought I'd chosen the wrong person, that it was really number four.

I was unsure, but it turned out that the police were not. They had plenty of evidence to revoke the parole of number four, who had murdered a woman years before. Someone who knew him from his job at *Street News*—a paper sold by homeless people on subways and street corners—and had recognized him in the composite pictures based on my identification had reported him to the police. He'd skipped two recent meetings with his parole officer, so officially they picked him up on a parole violation. When the new charges against him went to the grand jury, there was enough evidence to deny him bail until the trial—an event, I was told, which would likely not occur for another three years. Until then, because of the attacker's parole violation, he was sent back to prison to serve out the remainder of his sentence for the murder of another woman.

After the suspect was arrested and sent to prison, almost everyone expected me to breathe an audible sigh of relief and go on with my life. But no one knew the degree to which I was just pretending to be okay. I wasn't totally convinced who was in that jail cell, and I wouldn't have that peace of mind either way for many years, until the trial.

My parents lived on the Long Island Sound, and there was a long dock off a pier just behind their house. In the last days of that summer, I remember sitting on the dock cross-legged, wondering what would happen if I just went into the water and stayed there. I never had any urge to kill myself, but I wanted to be gone, just the way everything I'd ever believed in was now gone. I felt stuck in a life with no God, no sunshine, no light. I was twenty-two years old, the best part of my life was over, and I didn't know how I was ever going to be able to take a deep breath again.

In early September, just over three months after the attack, my father said to me, "Let's take a look at your résumé." Whatever was left inside me—the little pile of ash that had been holding me up since the attack—was blown away at that moment. I emotionally disintegrated.

I couldn't go to work—I couldn't even walk in the city. I still needed the trunk checked before I could get into the car. I still slept with the lights on. My stab wounds weren't even fully healed. Moments like that, when everyone around me was going back to normal and wanted me to do the same, I felt like I was existing in a netherworld of fear and victimhood. I had no idea how to get out, and I was heartbroken that the people around me couldn't see me down there, stuck in a hole with no ladder.

I'm sure my father thought he was helping me to get on with my life. Maybe he felt work would be a good distraction for me. Whatever his motivations were, it was my cue that it was time to leave. I thought, I'm on my own here, they don't get it. If I stayed, I would either totally lose my mind or push them all so far away that I'd really, truly be alone.

I could no longer stand being in my parents' house, and sinking into the Long Island Sound wasn't an option, so I decided to run away. My college boyfriend had moved to Boston, and although we'd broken up after graduation, we'd stayed in touch. He wasn't the most emotionally available man in the world, and that suited me just fine. Numbness was what I wanted. He found me a room in a house not far from where he was living on the outskirts of town.

One day I was scared to be alone, and the next I was driving off by myself to a strange city to live with people I didn't know. After spending three months as the stab victim, the girl in the newspaper, the daughter who needed to grow up and get on with it—it was time to disappear.

CHAPTER THREE

This Is Not My Fabulous Life

Marius and I started dating when I was a junior in college. He was in a frat that my sorority hung out with all the time, and my friends all dated his friends. So when I didn't have a date to my sorority formal I decided to invite Marius. We kind of fell in together after that—there were no fireworks, but it was easy and fun. He was good-looking, athletic, and universally popular, the kind of man that everyone wanted to be around. He was always funny and flirty, and I loved his hazel green eyes and his crazy long eyelashes. I adored his family, and stayed with them often—he had four siblings and I became close to all of them, and I was especially fond of his father, who was a grumpy surgeon on the outside and total mush on the inside.

Marius and his family were a huge part of my life for two years—which might as well have been a lifetime at that age—but we were never soul mates or passionately in love. He was a man's man who loved to play quarters, while I couldn't even stand the taste of beer. I never imagined a brilliant future for us, but he was a good person with a tender heart and I knew I could depend on him. We started out as friends, and we ended up the same way.

After the attack, with most of my closest friends an ocean away, Marius was the first life raft I could think of. He was living in Boston, and God bless him, he was at my side the very next day after my parents called him. When most people saw me they'd cry, while I quietly pretended to be fine. With Marius, I gave in to the tears, and he raged at what had been done to me. Maybe that was why I felt safe when I was with him. His anger made me feel sane. At the end of that long, terrible summer—when I just had to get away from my parents and their talk about résumés and moving on—it made a crazy kind of sense for me to run to Marius. I can only imagine the panic he must have felt when his scarred, traumatized ex-girlfriend showed up in Boston. He was agonizing about his own future, and here I was on his doorstep dragging my load of personal troubles behind me. Good times.

Boston is such a blur to me now that I can't really remember what I did all day while I was there. It was just a place to go—a place where I could be anonymous and just breathe. The house where I lived was a shingled four-story home on a shabby, nondescript street in the farthest reaches of the city. A bunch of students and recent grads rented rooms, as well as one cop who was studying for his law degree. I remember him really well, and how his presence gave me a little comfort. I rented the gabled room at the very top, and I still couldn't bear to sleep alone, so mostly I slept at Marius's. Sometimes he slept at my place. At some point I had sex with Marius for the first time after the attack, just to prove to myself that I could. I so badly wanted to feel normal, and to know for sure that the attacker hadn't stolen away any pleasure I could feel from being touched by another person. That first time, I cried my way through it. Poor Marius probably needed a few stiff drinks just to get through it himself.

Marius was biding his time working at a hotel parking cars, trying to figure out what he wanted to do with the rest of his life. Meanwhile, I scuttled around his existence waiting for him to come home—something that was so not *me* that even now I have to shake my head. I was just hiding and waiting, waiting and hiding. Given what I'd been through, no one would have blamed me if I'd fled to a silo in the middle of a cornfield in Iowa—but that wasn't any kind of life I'd ever imagined for myself. I was the girl who was going to take Manhattan by storm, who was going to wear hot suits and fierce shoes and go to the best parties. Instead, while Marius was at work, I returned to my sad little room in a house full of strangers.

I have a vague recollection that maybe I worked in a shop, but my overarching memory of that time is silent paralysis—me under that gabled roof trying to hold it together, my thoughts spinning and spinning with no place to rest. I had a few cousins living in Boston at the time, and I remember that I met them a couple of times for drinks on Newberry Street. They knew what had happened to me, but we didn't talk about it. By that point the dressings on my stab wounds were gone and the stitches were out, so I looked normal on the outside. The worst of my scars were tucked away under my clothing—hidden and unacknowledged, just like all the mess inside me.

Marius was incredibly kind to me, and I felt protected when I was with him. He had too much to deal with in his own life to help me with my problems. But he formed a shield around me and carved out a little space inside where I could safely and quietly lose my mind. With him, I could just exist for a while, free of intervention and the need to explain myself to anyone. Marius didn't give me anything emotionally, but he didn't ask for anything, either. My time in Boston got me away from the alternate universe of pretending that I now associated with home, and the constant noise of people talking at me, talking about me, stepping around me—compelling me to worry about their feelings and take care of them when what I really needed was someone to take care of me. When I fled to Boston, it was like I put on headphones. All external sound was blocked, and I could try to think again.

One gray day just before Thanksgiving, after I'd been in Boston for two months, I woke up alone in my room. Marius had left, and the house was quiet. It was a weekday, and I looked at the alarm clock and realized that everyone had gone to school or work. They were all off living their lives, but what was I doing?

I was just lying there, waiting for someone to walk through the door and make my life better. But no one was coming. And if I stayed in that tiny room, in that insulated, joyless existence, then I really had died that day in May. I certainly wasn't alive. I was miserable and empty, and my past was like a dream of some long-gone era. I imagined the old Jenny Gilbert—the girl who'd hitchhiked in Sicily and backpacked in Portugal. Where was that girl, and why had she given up? I had let the attacker frighten me out of the city I loved, and into a deep hole that I had dug for myself to hide in. This was *not* my fabulous life. This wasn't why I had fought so hard to survive that day. If I was going to be afraid to live, then that man might as well have killed me. I was lost and alone and scared, and I was headed down a road of not just *feeling* nothing, but *being* nothing.

I woke up for good that day, and I made a choice to stop hiding. I stopped looking back, and realized something that's been my mantra ever since: "You can't move forward if you're staring in the rearview mirror." It was time to go home and get a job and make my own way in the world.

But first I had to take all the stuff that was dragging me into that hole—all the fear, grief, sadness, and uncertainties—and put it away. I wouldn't pretend to be okay when I got home (or so I thought). I would *be* okay. I would bury that junk so deep, lock it away so tightly, that it would never get out. So I envisioned putting it all in a box—every thought, feeling, and experience I'd had during those dark days. I closed the box and taped it down. Then I carried it down a long corridor in my mind. I opened up the last door in that hallway, and went into a room with a closet. I opened the closet and put the box on the top shelf. Then I locked the closet door and bolted the door to the room. And I walked back down that hallway and instructed myself never to look back.

I found my own way of moving forward. It wasn't therapeutically sound or advisable, but it worked for me—at least for a while. Jenny was gone, but so was the dead girl who'd lived in that attic room. There was a new Jen in town, and I zipped her up like a suit, a layer of resilient armor that told the world that I was fine.

I went to see Marius, and I thanked him. I said, "I know you had no idea what to do, and I love you for just being my friend through this." Then I packed my car and I went home. I wasn't going to live there for long. I wasn't going to work in some nice Westchester suburb, on some safe, enclosed corporate campus. I was going to live and work in the city. My city. That monster hadn't killed me after all.

Even though the girl who dreamed of a fabulous New York life had been destroyed, the dream itself remained. My first task was to figure out who the new me was who could make that dream come true. I had to construct that new me from the top down; I had to imagine myself up there in the penthouse and try to figure out what kind of foundation I needed. And not only did I have to rebuild myself in such a way that I could have that dream again, but I had to rivet together my new personality without the slightest perceptible weakness. That was the only way I would be able to survive my fears and navigate through them without anyone on the outside being the wiser.

The result of my rebuilding was an assemblage of contradictions, all hidden beneath a shiny skin. I was a fearless fearful person. I was isolated but afraid to be alone. I was terrified of things that most people take for granted—especially sleep—but the stuff that others approach with trepidation didn't even faze me. New career choices, job interviews, selling, cold-calling—that was nothing to me. I knew what it was like to almost lose everything, so the day-to-day things that cause the average person anxiety? Please. What's the worst that could happen to me—the interviewer wouldn't hire me? The prospective client would hang up on me? This is not scary stuff.

For my very first interview after I got back from Boston, my parents offered to drive me into the city. Insanely, I refused. They can't drive me in forever, I told myself. If I was going to be the new person I'd invented in my head—if I was going to get a job, get out of my parents' house, and put behind me all memories of that summer—then I couldn't give in to even a moment of weakness. So I put on my interview suit, and I tucked inside my pocket the can of Mace that my father's secretary had given me. I rode the train and then the subway—the same trip I'd taken back when I was attacked—with my hand on that can of Mace the entire time.

And it wasn't just my first time riding the subway that was so awful. For years after, my chosen mode of transportation caused me white-knuckled fear. Yet I rode the subway at least six times a day, every day, forcing myself each time to swallow my terror. I felt physically ill waiting on the platform, and then riding the train. My heart pounded, I scanned every face. Every dark corner, every rider, was a potential attacker. It was like repeatedly throwing myself into a lion's den. But I did it over and over again.

Something happens to you as a woman after a man picks you, follows you, and beats you. You question everything about yourself, and what it is about you that might have caused this to happen. The label of "victim"—someone to whom bad things happen—penetrates every facet of who you are. *Embarrassment* isn't a deep enough word to encompass the feeling that the word *victim* gave me. *Shame* is almost dirty enough, but even that doesn't quite cut it. It was disgust; it was soul-sickness. To this day the very thought of being pitied makes my stomach turn in revulsion. More than anything, my desire to rise above this label is what propelled me onto the subway day after day after day.

Meanwhile I was haunted by "why" questions. Why had I been targeted? Why had I been spared? The answers to those questions tormented me. I'd been told that I was lucky to survive, but that made me feel that I wasn't inherently worthy of life—I had no special claim to simply exist. And if my life wasn't a right, I concluded that it must be a reward, something that had to be earned.

For years after, I would work harder than I'd ever worked before just to *prove* myself worthy to others—to bosses, to clients, to men, to the universe. I had to prove to my parents that I could pick myself up and get a job. I had to prove to the attacker that he hadn't won. I wrapped myself in so many layers of ballsy self-confidence that no one who didn't know better would ever guess the wells of shame that lay beneath. I was always on the defensive, always ready to counterattack: *What do you mean, I'm not worthy? Well, I'll show you.*

I lived on the energy output of my pure defiance. No longer was I the dimply, happy-go-lucky nice girl from the suburbs. No longer was I the messy procrastinator who flew by the seat of her pants. No longer did I take it as an article of faith that everything would be fine. The new Jen didn't take anything for granted. I was fierce and on top of things—I was type A-plus-plus. I faced down all opposition—the attacker and anyone else in the world who tried to tell me that I couldn't do everything that I set my mind to. I looked them all in the eye, I thrust my shoulders back, and I said, *You picked the wrong girl.*

PART II
Red Lipstick

We do not think our way into new action,
we act our way into new thought.

—ANONYMOUS

CHAPTER FOUR

Putting My Face On

Before the attack, when I imagined my grown-up life, it was always in big terms: successful career, handsome husband, huge family. After the attack, I lost any dreamy notion I'd once had of a storybook life. In contrast with my robust outward bravado, the future to me looked like an opaque dark cloud. So instead of looking at that gloomy horizon, I focused on the present and living in the moment. I decided that even if I could never again feel joy for myself, then at least I could immerse myself in other people's celebrations.

My first job interview was for an event-planning company that consisted of one man named Jonathan and an assistant in a windowless office downtown. Jonathan had a big wedding-planning business, and he wanted someone to build up his corporate clientele. He asked me how I felt about cold-calling. I said fine. He also asked me, "Where have you been the last six months?" So I matter-of-factly told him about the attack. I think it was at my second interview that I happened to catch sight of my résumé on his desk with the words "subway victim?" scrawled at the top. I cringed and my cheeks flushed hot when I saw that description of me in black and white. He hadn't made a note about my personality or my great attitude, just that I was a victim. Well, here was my first challenge, and the first person I would have to convince to believe in the new me.

Within weeks, I had two job offers—one from that small event-planning company, and one from the conference-planning department of Bear Stearns. Instead of choosing the nice salary at the big company, I accepted the job at the small firm. I loved that Jonathan was giving me carte blanche to build his business and that the results were all up to me. I loved that I'd be collecting a commission on any business I brought in. I loved that my boss handed me a phone book on my first day at work and said, "Go for it." I figured if I sold him on the new Jen after that second interview, then I could sell a party to anyone.

I worked nonstop. I spent the first months of the job learning every inch of the city. I was on and off the terror-inducing subway all day long, touring venues. Then I was back in the office making phone calls like crazy, and then crashing on Laura's sofa when I wasn't commuting home to Westchester.

When Laura went away for the summer, she let me stay on in her apartment, and my friend Deanna, just home from Italy, moved in while we looked for an apartment to share. I had met Deanna when I took an eight-week course in London after my sophomore year at the University of Vermont. I'd applied at the last minute through a New York University program, and I knew absolutely no one. The taxi dropped me off outside the London dorm along with my zillion suitcases (for an eight-week summer stay, let's not forget). I was alone, trying to carry all my bags, pulling one behind me and lugging more on each arm. I was probably wearing a hat at the time because I thought that's how the locals rolled. I got the very last available room in the dorm, a single that was so small I could touch two parallel walls at the same time (I have no idea where I thought I was going to put all those bags). I was struggling to get the door open when a gorgeous, exotic-looking woman with weird asymmetrical hair and multicolored nails emerged from her room. She looked at me with an amused expression on her face and said, "Can I help you?" She's been helping me ever since.

Incredibly bright, first-generation Italian Argentinian American, Deanna had grown up in Canarsie. After our first summer together, we traveled back and forth to our respective colleges to visit each other (I'd go to New York, she'd come to Vermont), and she always spent some portion of the summer with me. Having her move in with me at Laura's place was just a natural progression. Then, when my sister Rachel's roommate bailed on her at the last minute, Deanna and I found an apartment big enough for three. I'd finally found a way to repay Rachel for her kindness and for all those sleepless nights I'd put her through after the attack.

Everything was set, or so it seemed. I had my apartment, my roommates, my great job. I worked all day and went out every night, and then I was back at my desk at 8:30 a.m. the next morning with my cup of coffee and my phone book. I was a maniac. On some level I must have known that the more I kept moving, the less I had to think.

Thinking too much was never a good plan for me. That was when the darkness came flooding in, and the questions of "Why?" and "What if?" At those times—always when I was alone—I went into deep mourning for the old Jenny, the girl to whom nothing truly terrible had ever happened. On days like that, when I woke up feeling so awful that it was like my heart had been swallowed, my carefully constructed public face was all that got me through the day. I'd put on my most fabulous suit (with the sharpest shoulder pads), maybe even top it off with a hat. High-heeled shoes, always. Often I'd hot-roller my hair. The final touch, my armor, was red lipstick, my most aggressive shade. I remember it was back in the early days of MAC Cosmetics, when red lipstick was the new thing (again), and the name of the shade on the lipstick tube just spoke to me: Viva Glam. That's what I wanted to be, and when I wore that shade, that's what I tried to be. When I walked out in the morning dressed that way, Laura would say, "Uh-oh. It's one of those days, huh?" She knew the costume I was wearing was a sign that bad stuff was going on inside me. I couldn't fool her, but as long as the outside world didn't know, then I felt like I'd won a little battle. Instead of asking me how I was, people would take one look at my outfit and say, "You look nice, where are you going?" It was the trick of illusion—perception is reality.

The only ones who knew about the attack were my family, and my old college and high school friends. And I made absolutely sure that it stayed that way. It wasn't hard to keep it a secret. My name and picture had never appeared in the news reports, and this was twenty years ago, long before news was just one Google search away.

I never spoke of my attack, and to anyone new I met, I was exactly what I appeared to be—confident and on top of the world. But at night, when I took my mask off, I'd look in the mirror and cry. That was my own business. I could get to know everyone, form relationships, yet no one would ever really *know* me. Not in a million years would anyone guess that the pretty girl in the tight suit and heels had ever had a day of uncertainty—let alone that I had suffered a soul-destroying trauma.

My starting salary in my new job was $18,000 a year plus commission, and I didn't have the faintest idea what I was doing. I didn't know a plus-one from a plus-plus (the latter, I eventually found out, was tax and gratuity—who knew?). Jonathan had hired me to find new business. "New business" meant that the companies I'd be calling had never heard of my company, and had no interest in paying me to do what they already did themselves.

I asked myself: Who spends money? Bankers, lawyers, and accountants, of course. So I went after Bear Stearns, Morgan Stanley, Goldman Sachs, every bank in the city. This was 1992, and I remember at least one corporate type saying to me, "It's too bad you missed the eighties, honey." (Little did they know how the 1990s would eventually look like the good old days.) I called in-house conference planners, secretaries who worked for CEOs, anyone who had influence over corporate events. I wriggled my way in and then convinced them that they needed me—a twenty-three-year-old girl with no experience in the business, and who didn't even know how many people you could fit into a thousand-square-foot room. My chance of succeeding was slim to none, but after what I'd been through, it seemed like a cakewalk to me.

My typical workday those first few months went something like this: I sat at a desk staring at a brick wall in our windowless one-room office. My boss sat two feet away from me, and never offered a word of encouragement. I'd cold-call someone at a big corporate office and say that I wanted to plan their next event. They would hang up on me. So I would call back. "Hi, it's Jennifer Gilbert, your friendly stalker! I'm going to call you every day until you meet with me." I'd smile so hard that they could practically see my dimples through the phone. I was relentless, just like my dad had taught me to be. Every phone call was a challenge, another opportunity to prove that I was worthy of this life.

If that didn't work, I'd start sending them things. The corporate event planner for Salomon Brothers was a particularly hard nut to crack, so one day I got fifty yellow balloons, and I went to her building and sat in the reception area until she agreed to just listen to what I had to say. She became one of my first clients.

Still, it was rough going trying to convince cash-strapped corporate clients that they needed me. I knew that if I was going to really make inroads, then I had to come up with something substantive that my clients could never swing for themselves. I asked myself: what do my clients want that they don't already have? Whatever it was, I had to give it to them.

The answer was simple, really. People always want what feels just out of reach. Beautiful people want to feel smart, and smart people want to feel beautiful. Fashionable people want to be rich, and rich people want to be fashionable. Applying that math, I knew exactly what I could get for my corporate clients that they couldn't get for themselves: stylish connections. The really hot new clubs in the city didn't want a bunch of bankers as their clientele, no matter how much money they had. But those bankers very much wanted to be in the hot new clubs. Meanwhile, it's not as if I was born with a golden Rolodex of society contacts. I had to make my own. So I visited the hot places while they were still under construction, and I booked my clients into them before they became untouchable. By the time my clients' events rolled around, they were having their holiday parties at clubs so exclusive they never could have gotten into them on their own.

My determination to prove myself never slackened, not even for a second. By that point I was doing lots of bookings, and I had up to fifteen events on any given night. Every event felt like a make-it-or-break-it moment for me, but I remember in particular an event at Chaos. I'd done a walk-through of this club (soon to be SoHo's hottest spot) when it was still a concrete block, so I managed to book a venerable investment bank into the club for their Christmas party, which was a huge coup. The club convinced my client that she should use Chaos's DJ. I was wary—I only liked to hire my own people—but she thought it seemed like a good idea. We were both young, in our early twenties, and had a lot riding on this event being over-the-top amazing. But I was right to be concerned; the DJ who showed up that night might have been perfect for a bunch of club kids, but not for these investment bankers. They wanted Top 40, while the DJ, a snotty-looking kid with slicked-back hair, was playing nothing but trance. When the panicked corporate planner asked him to please play some dance music, the DJ said, "This *is* dance music." By this time her bosses were giving her dark looks, and she was on the verge of tears, certain that she'd be packing up her desk the next day.

I asked the DJ if he had any other music, but he had nothing, just a stack of vinyl that he was spinning himself. Now there were four hundred investment bankers standing around, complaining that the music sucked. My motto has always been, "A lot of booze and a packed dance floor makes for a great party." Well, this dance floor was a vacant wasteland in the center of the room. Not a soul dared to bust a move.

These were all millionaires who could afford their own bottle service and platters of thousand-dollar caviar if they'd really wanted a fabulous night. But that wasn't the point. They were masters of the universe, and if you threw a party for them, then they had expectations. The mood in the room was turning ugly, the planner was now genuinely in tears, and

I was staying just one breath of air ahead of my own bubbling panic attack. If I screwed up this job, if word got out that I was incompetent . . . why would any other client hire me ever again? It was as if my whole career—my life—was on the line. I felt a surge of adrenaline, all the tell-tale signs of fight or flight—pumping heartbeat, skin beginning to tingle.

Panic wasn't an option, so I got on the phone with someone in my office who was handling another event for me that night. That party was well under way, and I quickly hijacked one of those DJs, leaving his assistant to finish off that event. Then he and I grabbed every CD from our cars so he could start playing "Oh, What a Night" instead of *Club Mix Ibiza*. When the Four Seasons cranked up, the banker boys started dancing, and the crisis was averted.

That investment bank became a steady client, and the charge I got from being the fixer made me feel like I'd justified my existence yet again. My clients' needs were my needs, and their priorities were mine as well. I was proud of my skill at making other people happy. It was the only way I could feel happiness myself—a secondhand kind of pleasure.

I would do anything to make an event perfect. I was so adept at camouflaging my own flaws that I could also zero in on what wasn't exactly right in the world around me. Very quickly I acquired a practically supernatural ability to pick out the one tiny flaw in any room, no matter how large—the single wilted rose among dozens of flowers in the one centerpiece among hundreds. I'd walk right up to it, pluck it out, and everyone would look at me like I was Rain Man.

By the age of twenty-four I was already developing a reputation with the management of many of the city's largest hotels and catering venues. While most of my clients and colleagues were women, the directors of sales and of catering were still part of the old boys' club—but I was never intimidated. When I knew my client would be happier with a different price or menu, I got it for them. I remember one particular floor-plan standoff with the manager of an elite private club. He actually pointed his finger at me and said, "You listen here, young lady, I've been here for fifteen years, and this is where we put the bar." I smiled politely and said, "I'm the young lady who booked two million dollars of business at this club last year, and we are putting the bar over there." He moved the bar.

When people told me they thought my job sounded glamorous, I had to laugh. I loved it, but glamorous it was not. A half hour before the start of one of my big corporate events I would be down on my knees picking up gum from the carpet and sweeping up the flower stems that I didn't want the "fancies" to see (that's what I called my black-tie clients who booked the $1,000-per-person dinners). Once the event got under way, I still couldn't relax. Often enough I was sweating under my dress filling in behind the bar, pouring champagne so the guests wouldn't be frustrated by long lines.

This was around the time of my darling tulle bride's petticoat fiasco, and I was quickly learning that every event involved pulling some kind of magic trick out of a hat. That was where I excelled. It's not that I didn't get nervous or freaked out when something went wrong—especially in the early days, I could get just as panicked as any bride-to-be—it's just that I kept functioning when other people were busy losing their minds. So when we ran out of beer for a Salomon Brothers party at Tavern on the Green, I hopped in a minivan with the catering director and bought every available six-pack in the nearest supermarket. When everyone else was standing around gasping, I went to my fix-it place.

Pretty soon, I learned not only how to function in stressful moments, but also how to make those moments less stressful for everyone else involved. Events are magnets for disaster, and even I couldn't always make the problems magically disappear. So I had to figure out ways to help my clients handle their emotional responses. While the natural tendency in the face of man-made disaster is to escalate it even further by getting upset and pointing fingers, I had to gently, diplomatically help my clients see that while they might be justified in flipping out, it wasn't going to make the situation any better.

I handled the annual fund-raising auction for an upper-crust Manhattan private school whose auction committee was made up of parent volunteers (i.e., moms). They all had strong opinions, lists of absolute musts and must-nots, and they all had to agree. It took six months of planning and endless meetings, but I found the perfect party space for them that would meet all their requirements. Then, the day before the event, I received a call from the space, telling me that their liquor license hadn't come through. Now *this* was a disaster. Number one, it's a well-known fact that people bid more in auctions when they drink. No booze meant a lot less money for the school. Number two, all those parents weren't paying babysitters for a Friday night so they could go out and stand around drinking fruit punch.

Obviously I had to fix this problem. The women on the auction committee were up in arms. They wanted to sue the owner, they wanted all their money back. They probably would have put him to death by firing squad if they could have. But I knew that none of these responses was going to get them what they really needed: a successful auction. So I called the owner of the space, and instead of threatening him with some kind of nuclear option (which I knew would only result in him hanging up on me), I said, "Dude, you're killing me. Help me out here." In the end, he worked out a deal with another space, and he put on the event for half the original price. In addition, I got him to pay for shuttle buses for any of the parents who accidentally showed up at the wrong location.

Did the new space fit all those nonnegotiable requirements that the auction committee had set in stone for me? Absolutely not. But the space was available, and it had a liquor license, and the school made a whole lot more money than they would have had they gone to court. In less than twenty-four hours we had a solution that made everyone happy, and the school was so delighted that they hired me to do the auction again the next year.

I conduct my business by the guiding principle that people are people. Everyone wants to be treated with respect—whether they're responsible for the mistake or they're the injured party. My auction ladies deserved to be upset, so I let them have their anger—but I made sure it wasn't indulged on a phone call to the owner of that space. I got them to take their deep breaths, and then I made the call—because I knew that I'd get a lot more by being nice than I ever would have gotten by screaming his ear off or making threats. Sometimes, just because you can, doesn't mean you should.

Even when the stakes were a lot lower than a missing liquor license, I never dismissed a client's feelings. When a client had a nervous breakdown because the tissue paper in her invitation was the wrong shade of blue, I didn't try to argue that it really didn't matter. I didn't try to make the client feel selfish or unreasonable. I didn't tell her, "Well, at least the printer spelled your name right." I certainly didn't tell her that there were bigger problems in the world than colored tissue paper. I had learned the hard way that everything in life is relative, and that perspective only comes *after* the crisis. All those times that my own pain had been framed in relation to someone else's (*At least you weren't raped*) had taught me a powerful lesson: Never "at least" someone else's pain away. Let them have it, feel it, and then try to alleviate it.

CHAPTER FIVE

Running as Fast as I Can

Shortly after I moved back to the city, I became obsessed with running. I'd always been athletic, but I'd never been the kind of person who exercised in a disciplined way. That changed after the attack. It became very important to me to be strong, and to be able to run far and fast. I ran on city streets, in the park, in the gym, in the Hamptons. Every day, no matter the weather, I ran, and I always ran alone. Exercise was not social to me; it was my private time, where I built my strength and my resolve. While I was running, I was all eyes and ears—the two senses that had failed me the day of the attack. I constantly scanned my surroundings, my radar always on high alert for any threat. This was the new me.

While I ran, the naive, giggly, silly girl that I used to be sometimes came back to me, and I allowed myself to remember her. She was a nice girl, and I missed her. But bad things happened to that girl. This girl—the new me—stayed on guard, armored and ready.

Anyone who has experienced a heartbreak or a devastating loss knows the feeling of channeling emotional pain into something physical and tangible. I funneled all of my pain into action. I was always on the move. It was the law of inertia. A body in motion stays in motion.

I worked constantly, never taking a vacation, and when I wasn't work-ing then I was working at playing. In the summer I took a weekend share in a Hamptons beach house that was so insanely packed that you had to get there early and guard your bed. After a weekend of socializing, on Sunday night I'd drive back to Manhattan and head straight to Boom, a chic, funky place in SoHo. They reserved a table just for me—not to sit at, but to dance on. I was on every list of every party, every junior com-mittee, and all those events showed up in the *New York Times* Style sec-tion party pictures. Because I booked their spaces every day, I knew everyone at all the clubs, from the bouncers and the coat-check girls to the hostesses and the bartenders. At night, when I had a few girlfriends in tow, I'd smile to the bouncer and he'd wave us in, and then I'd wave to the bartender and he'd have my drink ready. It was like a magic trick, and each time I couldn't believe I'd pulled it off. I'd wanted a fabulous life, and here it was all around me.

Marius was my one serious boyfriend for most of college, but now I found myself surrounded by single men. All my friends were at the pairing-up stage and starting to think about husbands and future fami-lies. But I was emotionally numb. When my friends talked about their dreams, I was happy for them of course, but on a basic level I couldn't relate. All of these romantic fairy-tale thoughts that I had once had were gone. The happily-ever-after felt impossible to me.

Still, every client seemed to have someone they wanted to fix me up with, and I became queen of the blind date. In a weird way, I was more outgoing and vivacious after the attack than I'd been before. I was game to meet anyone, and I could make conversation with anyone. Deanna would joke that I could chat up a fruit fly. On one record-breaking night I actually had three dates in a row—one for drinks, one for dinner, and one for dancing. When a date arrived to pick me up, the doorman would call up, and I would come down to meet him. We'd have to go on at least several dates before I'd even invite a man up to my apartment.

I wasn't playing hard to get—I genuinely didn't want to be gotten. Not only did I feel so emotionally empty that I had nothing to give in return, but I dreaded the moment of truth when I'd have to tell a new man about the attack. I still had scars in so many hidden places that I couldn't possibly have sex with anyone without revealing my secret. So I didn't get to that point with many men, at least not for a good long while. When I did become intimate with someone, I insisted on keeping the lights off. But the ironic downside of keeping my scars so well hidden was that it made the inevitable unveiling loom all the more dreadful in my mind. I lived in fear of a man noticing my tic-tac-toe board of scars. They were my scarlet letter, an indelible sign that something horrible had happened to me. They marked me as a subject of pity, and brought up all those feelings of shame that I kept so carefully buried. It was much easier for me to keep men at arm's length than to have to deal with all of that.

I sometimes wondered if it would be better—healthier—if my scars were more visible. People had told me so many times how lucky I was that the attacker hadn't gotten my face, and of course I knew that. But on some level I thought, If only people could see my pain just by looking at me. Instead, my scars were hidden away under my hair, under my clothes, fading to whiteness on my hands and between my fingers. For anyone to know what I had gone through, and what I was feeling inside, they'd have to ask me. But as it was no one ever suspected—or at least they never asked. So this breathtaking disconnect continued between how I appeared and behaved, and what I actually felt like inside. I was a gregarious lonely person, a party planner celebrating other people's amazing life experiences, and I was running as fast as I could away from my own problems, acting for all the world as if I didn't have any.

Jimmy was the first man after the attack that I allowed myself to love. He was a friend of one of my very good college friends, so he was not a random stranger, and somehow that felt safer to me.

Everything about Jimmy was welcoming, and miraculously, I relaxed into him as into a warm embrace. He had brown hair, a great smile, and blue eyes with an eternal twinkle. He was the rare "guy's guy" who actually loved to dance. Dancing always felt so freeing to me, I could get lost in the music, and he found me there. He was funny and outgoing, but also incredibly sweet, and infinitely kind. Whenever he rang my bell, I'd open the door to find him standing there with my favorite malt balls in one hand and gummy candies in the other.

After we got together, we stayed together—every night, weekend, holiday, vacation. I was twenty-five, Jimmy was twenty-seven, and neither of us was thinking about marriage at that point—least of all me, since I was still focused on getting through each day. But in every other way our lives were completely knit together. Even the nights we didn't plan to spend together, chances were he'd show up on my doorstep at 2:00 a.m. He held me when I woke up with nightmares, and he knew every one of my scars, inside and out.

Jimmy saw me through the most difficult years of my life; he was the first new man in my life I told my whole story to, and he signed on for everything. He was the definition of a really good man, the best of the best, and he'd do anything for me. Just one example: my family planned a Cayman Islands vacation, and Jimmy was invited along. My father is a scuba-diving fanatic, so for every single one of his holiday gifts, my relentless father gave Jimmy a flipper, a pair of goggles, a snorkel. Jimmy smiled along with each gift, but the truth was that he had never snorkeled or dived—he didn't even like to swim. Still, he never let on. Instead he spent two days of his Cayman Islands vacation taking the resort's scuba-diving course, just to make my father happy—because he knew that would make me happy.

Jimmy was right beside me on my dad's sailboat when I first hatched my plan to start my own business. I'd had an entrepreneurial streak since I was eight (the first time I'd charged admission to one of my backyard circuses), so it was never going to be enough for me to take a salary and earn a commission. I wanted to own something. Sure there were other people out there planning events, but I wanted to do it for myself, and do it better.

I'd been working for the same event-planning company for just over two years, but already I was responsible for half the company's revenue. I was the one breaking down those corporate doors convincing companies to hire me for their events, creating customer loyalty, getting repeat clients. In addition, I had come up with a whole new way of approaching the business of event planning.

I'd expanded my boss's business by pairing up clients with great venues, but I'd also heard the same complaint over and over again. The venues were desperate for corporate business, but my corporate clients complained that they couldn't afford to pay me a retainer. The corporate event planner didn't have the time to find the newest, most perfect venue, and they were desperate for advice, but they didn't want the higher-ups to know that they had hired someone else to do their jobs. So they just kept booking the same spaces for the sake of expediency.

In a eureka moment, realizing that my industry had been going at this thing all wrong, I flipped the business model. It wasn't the businesses that should be paying the event planners—it was the venues. So I became a PR and marketing agency for any sort of event location. Of course the companies were thrilled, because this meant that I would provide my planning services for free. And the venues were thrilled because suddenly they had all this new business. Within months I had hundreds of spaces wanting to sign up to be in my database. This idea was revolutionary, but it didn't occur to me to pat myself on the back. I figured I was just solving a problem.

Around this time, one of my biggest clients, owner of one of the most popular clubs in the city, offered me a job as his full-time event booker. I told him it was a great offer, but I didn't want to just bring him business—I wanted to do business with everyone. I'd get bored just booking events for one space, when what I really wanted was to build my own company. I was so committed to my plan that I'd already named the company in my head—Save the Date. It never occurred to me to name my own company Jennifer Gilbert Productions, or anything with my name in it. These events I planned were not about me; they were about the clients, and what they wanted their own event style to be. I was just the person who figured out how to make it happen for them.

But that club owner just wouldn't take no for an answer, and finally we came to a brilliant compromise—he'd become my partner in my new company. He'd pay my overhead and give me office space. I'd book all of his events, and I'd also book events at other venues throughout the city. It was a win for both of us. He made money no matter where I booked, and I achieved my dream of running my own company, but without the headache of managing an office.

I left my job, paid my old boss the commissions on all the clients I took with me, and became totally, joyfully consumed with building Save the Date®. I had put away all those old fantasies of a blissful future of having it all—but maybe the universe would let me have this one thing, my bouncing baby company.

CHAPTER SIX

Chaos Theory

I knew the district attorney expected that it would take three years to bring my case to trial, but somehow I managed to put that knowledge on a shelf and pretend it didn't exist. For three years after the attack, I didn't have any conscious sense that a monumentally traumatic event was hovering over the horizon right in front of me. Within me, though, there was a timer ticking away. Little did I know that when that timer went off, a world of hurt would explode.

Whether it was coincidence or not, in the months prior to that big impending milestone, I threw my life into an uproar. Looking back on it now, it's hard to believe I wasn't doing it on purpose—the timing was just too perfect, and the results were guaranteed to draw my attention away from my immense fear of facing the attacker.

First, I started a company that swallowed me whole and gave me no time to think or pause. Then, as if that wasn't enough to overwhelm me, I broke up with Jimmy. The launch of my company was a welcome distraction—even if the timing was a little ill-advised. But breaking up with Jimmy at that moment was so insane that it was self-destructive.

I told myself that Jimmy and I weren't really meant for each other. Travel was my passion, and he'd never left the country. He liked football, and I hated it. We wanted different things from work and life. None of those reasons for breaking up with him were wrong per se, but they weren't the real catalyst for me choosing that exact moment to end what had been a perfectly happy relationship. In truth, the rational me went into high gear and wanted to protect Jimmy from the chaos that was about to ensue.

Falling in love with Jimmy, I'd let my guard down. He allowed me to be vulnerable, and in return I didn't put up any walls with him. But the soft, needy person that I could be with Jimmy would never survive what I was about to endure. So I walked away from that sweet, steady man and went back to work constructing my walls.

It's of course deeply ironic that while I was running away from my own long-term relationship, it was my job to plan weddings for other young couples. I could always be far more optimistic on someone else's behalf than I could for myself.

Planning someone's event, particularly a wedding, is a very intimate and complicated process. Most couples don't know exactly what they want when they meet with me, and it's my job to figure out what their dream feels like, or looks like. Sometimes all I get is a song they love or a color. One couple said, "We feel strongly that it's all about blue." Now I'm thinking, Blue like a Smurf-themed wedding, or blue lighting, or simply blue hydrangeas on the tables? It's my job to figure out what they really mean. Sometimes I know what they want by eliminating what they *don't* want, so I can create some ideas from the other end of the spectrum. A couple might say, "We want something totally different," but when I show them a raw industrial loft space they say, "Oh, not *that* different." So then I realize that to them, "different" just means that they don't want it in a hotel ballroom. It's all a matter of asking the right questions, and all of us being open to the discovery. I become part detective and part therapist. By letting people reveal their fears, desires, and needs to me, I can piece together their perfect event. This all seemed perfectly logical to me, except for the fact that I was the one in desperate need of that same self-examination and therapy.

Of course there are times that even my on-the-job event therapy still isn't enough for some persnickety clients. I remember one particular wedding when the bride actually locked herself in a bathroom not once but *twice* during the reception. Each time she threatened not to come out, and each time I had to lure her back to her own wedding by painting a happy picture of her married bliss.

She was the perfect example of what I've come to call binder brides. They're the women who walk into our first meeting to plan their wedding with a loose-leaf binder stuffed full of pictures and ideas that they've been collecting since the age of five. In my experience, these women are almost impossible to please—no wedding can ever live up to their outsize expectations. They forget about the meaning of their upcoming event, including their fiancé's feelings, because for them it's all about the fantasy they created years ago. Usually I'd figure out a way not to work with brides like that, but sometimes I went back on my own rules. This particular bride, Bunny (her given name was Madeline), had come to our first meeting with everything she wanted for her wedding already booked (and neatly filed away in her binder, of course). She'd picked the florist and the band and reserved the Central Park Boathouse months before she even got my name as a planner. The only thing she wanted me to do was coordinate the day of, with neither my input nor my opinions.

The wedding was set for the beginning of September, and back then the Boathouse wasn't air-conditioned. I strongly advised the bride about one hundred times to bring in portable air conditioners, because September in Manhattan can be hot and humid. But her father was watching the budget, and he insisted my suggestion was unnecessary. A hot reception is a miserable reception, and so I begged them at least to pay a deposit to reserve the air conditioners—then they could decide a few days before if they'd need them. Still the answer was NO.

Inevitably, the wedding reception was a disaster from beginning to end. The first time Bunny fled to the bathroom, it was at the sight of the centerpieces. They were supposed to be huge topiaries in the shape of bears (to me they looked like Edward Scissorhands meets *The Three Little*

Bears). Bunny called her fiancé Wade her "little bear" (they were a match made in Noah's Ark) and thought that would be a great idea for a floral theme. Bear centerpieces? Sort of random, but to each her own. Unfortunately, Bunny thought that the topiaries came out looking more like monkeys, sort of bears but with oddly shaped tails, and she was apoplectic. She ran into the bathroom in a rage, screaming, "It's not at all like the photo I showed my florist!" Then she locked herself in a stall.

I could have just thrown up my hands—after all, the bride's nervous breakdown over floral arrangements she had chosen way before she chose me wasn't really my responsibility. But the secret, hidden romantic inside of me couldn't bear to see a beautiful wedding ruined over something that couldn't be changed at that point. So while she cried, I told her how lovely the wedding was, and how lucky she was to have found the man she wanted to spend the rest of her life with. I said what a miracle it was that everyone she loved was just outside that door ready to celebrate with her. When she continued to rail about the monkey/bear topiaries, I said, "Bunny, *honey*, they are on the tables and your guests are all seated. No one knows what they were supposed to look like but you, so you need to move on now and enjoy your wedding!" I said she could call the florist and rant after the honeymoon, but right now was not the time. Finally, she stopped crying and came out. But that's not the end of the story.

So of course it was a gorgeous September day that hit eighty degrees. We were all warm, but she had on this enormous dress—her dream dress, but the last dress you'd want to wear on a Saturday afternoon in a park with no air-conditioning. I tell every client, "Practical is beautiful." But that advice wasn't going to help this client, not now. There she was, in layers and layers of heavy and HOT satin. After the first dance set she was dripping, and her makeup was a mess. She was already emotional, and the heat pushed her over the edge. I saw another breakdown coming, and I said, "Let's get you into the ladies' room and clean you up."

While I was getting out her blow dryer, she was looking through her makeup bag and she had to pee, so she unzipped her huge dress and went into the stall with her bag. Somehow she thought it would be a great idea to pour baby powder all over her body, and instead of soaking up her sweat it had combined with the sweat to form a thick white paste (think cement or papier-mâché). When she emerged from the stall, she looked more like a kabuki stripper than a bride, and then she started screaming at me, "GOD DAMN IT, HELP ME!" We were both in a complete panic. She tried to pull her dress back on, but her skin was so sticky that the satin wouldn't budge. Before I could tell her not to force it, she yanked and we both heard a rip, and her shrieking rose a good few octaves. I took her by the shoulders and said, "Stop! Take the whole thing off and *back away from the dress*."

Bunny was a full-figured woman, and was standing in her blue thong (I guess that was the something blue) in front of me and freaking out. Since we'd been away from the reception a good ten minutes and the speeches were scheduled to start, everyone no doubt would be looking for the bride. I told her, "This is your day, nothing starts without the bride." This was a bit of a crisis, and first things first. It was humid, a zillion degrees in that little bathroom, and we had to get her dress back on. I had to apply wet towels to scrape the paste off, then blow-dry her body, then get her (painfully) back into the dress, fix her hair, and send her back to the crowd. She looked at me and said, "Everything's ruined, my whole wedding is ruined!"

Once again I told her how loved she was, and how sometimes things go wrong, but it would be okay—only I knew what was happening, and the love of her life was outside waiting for her. This time, though, I could see that my words weren't hitting their mark. I'd lost her. She stopped crying, but instead of going back to her reception with a smile on her face, she looked grim. There was no way anything was going to meet her expectations, and instead of just enjoying what she did have, she had a bit of a scowl on her face for the rest of the day. I forever thought to myself, Beware of the binder bride; you'll never make her happy.

I learned an interesting lesson from Bunny and her wedding. I saw the shadow side of having fixed expectations. They are an awfully hard thing to live up to. If you spend your time measuring your reality against your fantasy, you're inevitably going to lose the joy of just being in that moment. This holds true for events, relationships, business, and life. Bunny taught me that some people just don't know how to be happy, and they will never look at the bright side. And that's not a person I want around me in any situation.

While I was building my company and ending my relationship of two years, I was also deeply involved in every detail of Nicolette's upcoming wedding. Nicolette and I had met on my first day of freshman year at UVM. There was some sort of magnetic pull that brought us together, and we were inseparable. We'd grown as close as family in the three years that we lived together at college. She was a year ahead of me in school, and during that time she became the slightly older sister I'd never had. We used to laugh and say, "She's my sister from another mother." She was outspoken and confident, and I learned so much from being around her. While I was from the suburbs and relatively naive, she was sophisticated and worldly, the only daughter of a close-knit, loving Greek family that lived in the Bahamas. I spent most holidays with her family there, and they all opened their arms to me.

Nicolette's wedding would have been momentous for me anyway—here was one of my dearest friends embarking on a new life we had always dreamed about—and on top of it she had chosen me as her maid of honor. She had a six-pack of friends of her own and was extremely close with them, but she had asked me to be the only attendant in her side of the wedding party; it was just me and a mass of groomsmen. I was touched and honored, while also feeling no small degree of pressure to perform. Nicolette's fiancé, Hans, came from a prominent Norwegian family, so the wedding was going to be a spectacular international event—we only half jokingly called it "the coronation." Her soon-to-be in-laws actually began referring to me as Nicolette's lady-in-waiting.

A serious breakup, a new company, and your best friend's wedding are enough to send anyone in for counseling, but then that little time bomb inside me blew up. I was alone in the apartment I shared with Deanna and Rachel when the phone rang. It was the district attorney's office, telling me that the trial had been scheduled to start just days after I returned from the Bahamas. When Rachel walked in a few minutes later, she found me on the kitchen floor. I was having such a powerful physiological reaction to the news that I wasn't just crying, I'd lost all feeling in my legs. I couldn't stand up.

I went into therapy just to try to prepare myself to talk about the attack on the witness stand, and to emotionally survive being in the same room with the attacker. I found it so difficult to talk about the trial with the therapist, a woman named Ann, that it took me two full sessions before I could bring myself to tell her the real reason I was feeling a little stressed out. And if I couldn't tell her, how could I ever tell a jury? If I failed to convince the judge and the jury what he'd done to me, and the attacker was acquitted . . . oh my God, what then? The terror I'd felt while standing on the other side of that one-way mirror on the day of the lineup came back to me in waves of sheer panic. After I was finally able to explain what I was going through to Ann, we then began emergency therapy, and we met three times a week.

It didn't hit me until the phone call from the DA's office that I'd been in posttraumatic stress and denial for three years. How could I have achieved any kind of real closure since the attack, while knowing that the trial loomed ahead of me, and not knowing what the outcome might be? I'd managed to put a semblance of a life together—at least superficially. But internally I had never dealt with the attack, and that call propelled me into what I can only describe as a nervous breakdown.

And yet in the two weeks before the trial, I wasn't at home meditating and pulling my thoughts together, I was in the Bahamas with Nicolette, embarking on a dizzying number of wedding-week events. There were parties upon parties, and endless details. It was an exciting, highly effective distraction, enabling me to keep busy doing instead of thinking. In between tearful heartfelt speeches, outfit changes, and fits of giddy disbelief with Nicolette, I would suddenly feel a pang of terror. I had this secret internal countdown to the trial, and I couldn't talk about it to anyone. Who wanted to be Debbie Downer at this unbelievable wedding? So on went my red lipstick, and I concentrated on my friend's special day.

There was one unexpected oasis during that adrenaline-fueled week that gave me solace. My parents had been invited to the wedding, and one day my father came back from scuba diving describing the most magical place he'd ever seen. For my father to rave like that, I knew it had to be heaven, because in our sailing days we had seen every pristine beach that part of the world had to offer. He said it was a tiny little island two hours from Nassau by boat. There wasn't time for me to see it myself, but I kept an image of that little island in my mind, and it became a vision of peace in the midst of the chaos, soothing me in the weeks that followed.

By the time I got back to New York, even I had to admit that I needed more help than I'd been letting on. My therapist was guiding me through coping with testimony, but meanwhile I ached for Jimmy in a profound way.

When it finally hit me what a mistake I'd made—at least in my lousy sense of timing—I did my best to explain things to him. Miraculously, he took me back. He'd been devastated and uncomprehending when I'd broken up with him, and I couldn't expect him to fully trust me again. But he was too big a person to abandon me during the trial. He was one of the few who truly understood the road I'd traveled to get to this point, and what this next leg in the journey would take out of me.

CHAPTER SEVEN

Miss Gilbert, This Is Going to Be Very Hard for You

I lost about fifteen pounds in the weeks leading up to Nicolette's wedding and the trial. I lived on nothing but air, panic, and sheer force of will. While I wore all my hats of maid of honor, friend, and boss, the secret, singular focus of my life was to do my job at the trial: show up, tell my story, and finally see if the man who had been locked away all these years was indeed my attacker.

When I heard that the charge in my case was attempted murder, my reaction was both shock and vindication. It was horrifying to hear the words that said someone had genuinely wanted to kill me, and yet it felt like a kind of proof that I wasn't crazy. It was scary, but definitive. Now I knew for sure: I hadn't just survived an attack, I'd survived death. It might seem obvious to anyone who heard what happened to me, but the black-and-white terms of the charge against him were a revelation to me. No longer could I even try to downplay what had happened.

By the time my trial was scheduled, the alleged attacker had already been tried for his assault of the first woman, but it had ended in a hung jury. That attack took place on a set of subway steps, and it should have been a straightforward case of assault, but just one juror held out, and they couldn't reach a decision. The third victim had been so damaged—

I'd heard there were broken bones in her face and possibly even severe injuries to her eyes—that she hadn't been able to identify her attacker. This meant that I was now the only remaining victim who could say with certainty that he was the man who had done these horrible things.

The burden to put him away was all on my case, and I felt enormous pressure. Meanwhile, the last time I'd seen the attacker was through a one-way mirror, and I'd been so terrified that I'd actually identified the wrong man. As soon as I walked out of the lineup, I knew I'd made a mistake, and although I tried to correct it, I was afraid it would come back to haunt me during the trial. I dreaded being asked about the lineup, and the fear that I'd done irreparable damage to the case kept me up at night and closed my throat when I tried to eat. I was frightened of failing, and even more terrified at the thought of the attacker being acquitted and once again walking the streets—while knowing my name and everything about me.

These are some of the things I'd learn after I'd testified.

They called the attacker Mr. Clean in prison because he had a thing about keeping his clothes clean, and he obsessively washed his hands until they were raw.

Prior to the lineup, he had been identified from the posters all over town by someone he worked with at *Street News*. The first place the police looked for him was at the Laundromat where he always cleaned his clothes.

They finally caught him in Penn Station with a ticket to Boston in his hand. Not knowing this particular detail, I'd headed to Boston myself just a few weeks later.

It was early June 1994 when the jury trial began. It took place in one of the large, formal courtrooms in the New York City Criminal Courthouse at 100 Centre Street. The lead prosecutor for the Manhattan district attorney's office was a woman named Margaret Finerty. Tall and attractive, probably somewhere in her early to mid-forties, she came across as focused and aggressive. She was sympathetic but she didn't try to get emotionally involved with me—she was very much my image of the cut-and-dried all-business trial lawyer.

My strongest memory of the trial—other than my actual testimony—is of almost unbearable suspense. Because I was a witness as well as a victim, I wasn't allowed to sit in the courtroom and listen. The case lasted for weeks, and the wait for my call to testify was agonizing.

When I was finally called in to court, I waited again, gripped with nausea-inducing anxiety. Then I was told to go back home and come back the next day, because there wasn't time for my testimony. I felt like I had some sense of what it must be like to receive a death-row reprieve, but more than relieved, I felt overwhelmed and desperate. I'd already eaten my last meal, and I wasn't sure I could go through it again. I'd had to get myself to a very specific mental place where I felt capable of talking about the attack in front of a judge, jurors, and the attacker himself, *that day.* Now I had to step back from that awful place and convince myself that I could do the whole thing the *next day.* I didn't know how I could physically endure it.

It's surreal to remember that while I was barely controlling my fear of the trial, I was also doing my day job. Neither my staff nor any of my clients ever knew I'd been attacked, much less that I was about to testify in a trial. So while I was trying to hold down my own urge to throw up the contents of my stomach, I was also making calls to get one of my clients an emergency fitting at Kleinfeld because she'd lost another three pounds and was afraid her wedding dress wouldn't fit. I had to smile and reassure a client who was afraid that only 300 guests would show up for her event, instead of the 325 she'd counted on. I was simultaneously booking Christmas parties, choosing hors d'oeuvres, and managing my clients' freak-outs over their goodie bags.

Meanwhile, I was terrified that I wouldn't recognize the attacker in the courtroom, and I was even more afraid that I would. The viciousness of his attack against me, which had grown dreamlike in the intervening few years, became, once again, very real. I remember sitting on a bench outside the courtroom, certain that I was going to vomit. Detective Lontini—a warm, serious-minded plainclothes officer—sat with me, speaking calm, reassuring words. Otherwise, I was alone.

Originally, my father was the only member of my family who planned to attend the trial when I testified. My mother had insisted that she couldn't possibly attend, it was just too much for her to bear, and my father said he wouldn't force her or my sisters to go. When I told my therapist, Ann, that just my dad was coming to the trial, she stared at me in disbelief and said, "How does that make you feel?" I told her that I was relieved my dad was coming—that having someone from my family there was better than no one.

Ann told me that it just wasn't acceptable. Family shows up for the good stuff, and even more so for the really bad stuff. She said that it was time to ask for what I needed from my parents after those years of silence following my attack. She gave me the words, because I was so scared and uncomfortable that I didn't even know how to say it. I had to write them down for our family meeting after my session. I said, "No one wants to avoid this more than me, but if I have to be there, then you all have to be there." I said, "I need you all there, for ME." My family listened, so on the day that I was to testify, both of my parents were in the courtroom. Jimmy was in the courtroom as well, along with Laura, Deanna, and my sisters. The members of my whole family, related or chosen, were all there in a row.

Sitting outside that courtroom, waiting to testify, this is what I said to myself: no matter what happened in there, no matter what I was asked, I would not cry. And if I did recognize him, I would not shed one single tear in front of him. I would go in there and put that bastard in prison for a long, long time.

I was twenty-five years old, but the enormous mental and physical pressure on me made me feel a hundred. Other women my age were making their youthful mistakes, getting drunk in bars, doing their walks of shame the next morning. Meanwhile I was starting a company and testifying in an attempted murder trial. My heart pounded, my mouth was dry, my pits were wet. The bailiff came out and told me it was time.

Peggy Finerty had told me during our only meeting before the trial started that when I walked in, I should just stay focused on her. But I couldn't stop myself from looking at the attacker. He was right there, in the same room as me, with nothing but air between us. He was facing straight ahead and drawing or writing on a pad of paper. He didn't turn to look at me, but I recall so vividly seeing the back of his head and his shoulders—the same view I'd had when he'd straddled me and tried to stab between my legs—and a feeling of sureness jolted my body like electricity. It was him. I knew with one hundred percent certainty that he was the man who had tried to kill me.

When I got up on the stand, he was sitting at the table to my left. So I turned my entire body toward the jury, and raised one hand to the left side of my face in an attempt to block him from even my peripheral vision.

Finerty said, "Is the man who attacked you in this room?" I said that he was. She said, "Can you point him out?" I turned and pointed to him, and that was the one and only time that I looked him in the face. He looked exactly the same, maybe a little bit bigger.

Finerty then began asking questions about exactly what had happened on that day. At one point I noticed the bailiff handing something to the jurors, and the jurors passing it around. I remember the men shaking their heads, and a few of the women crying. I wasn't sure what was going on, and I felt confused, but I tried to stay focused on my answers to her questions. My mantra was, "Stay calm and breathe."

Finerty placed a stack of photos in front of me. She said, "Miss Gilbert, this is going to be very hard for you, but I need you to look at these pictures." I was the subject of all of the pictures. They were the Polaroids my mother had taken of me shortly after the attack. Then she said, "I hate to do this, but could you please identify these pictures?"

I felt myself getting emotional. Seeing those forgotten photos stunned me for a few minutes, and I remember the judge asking me if I needed a break. I said no. I told myself that if I could just keep it together and get my testimony over with, then I could get out of there and collapse where no one could see me. I identified myself in the pictures—although to this day I've blocked out all memory of what I looked like in them—and Finerty ended her questioning.

Then the defense lawyer stood up, and I remember looking at him and just hating him. The first thing he asked was, "Is it true that you didn't identify my client in the police lineup, and that you identified a different man?" I thought, Here we go. Of course he was going to ask me about this—it's what I had dreaded all these years. He gave me no chance to explain; all I could answer was that yes, it was true. He asked me more questions: How did you get a good look at the attacker? How did you know it was him if you were running? How did you know it was him if you were so emotional? How do you know it's my client after three long years?

I forced myself not to cry. I was not going to give his client the satisfaction of seeing that he'd broken me. I went to the calm place where I go when things get really awful. I couldn't even look at Jimmy or my parents for comfort, because they were sitting on the same side of the courtroom as the attacker.

When Finerty got up for her redirect, she asked me just one question: "Why did you identify the wrong man in the lineup?" I looked at her and then at the jury, and I said that I knew I'd picked out the wrong man. I told them how terrified I'd been, separated from the lineup by just a one-way mirror. But now, I had not a doubt in my own heart that the man sitting at the defense table was the one who'd tried to kill me. And backing up my certainty was that the composite drawing created from my description of his features was as accurate as a photograph.

After I testified, I walked out of the doors and fell apart. My vision tunneled, and I nearly fainted. Jimmy had slipped out of the courtroom and was holding me up. Then the bailiff came out and told me that they had another question for me and they needed me back in the courtroom.

All the calm that I'd saved up to get me through testifying had been spent. I had absolutely nothing left. My entire body was shaking—I felt like I'd just been in a car accident. My knees wouldn't hold me. I told him that there was no way that I could physically walk across that courtroom and get back on the witness stand. I couldn't be in the same room with the attacker again; I begged him not to force me. He went back in to have a sidebar conference with the judge and attorneys.

The courtroom was still full, and the attacker still sat at the defense table. They agreed to let me stand at the very back of the court, directly opposite the judge, and speak across the entire room instead of having to walk up the aisle. I would be shielded by the audience of people from having to see the attacker again. So I compelled myself back in there, and the judge asked me a follow-up question. He said, "Miss Gilbert, have you heard any testimonies or has anyone spoken to you about these proceedings? Remember, you are still under oath." I looked straight at him and calmly, honestly answered, "No, Your Honor, I have not." When I left the courtroom, Detective Lontini hugged me and Peggy Finerty shook my hand.

During the weeks of testimony and the wait for the jury to reach their verdict, I desperately wanted to know what was happening, and yet my body rebelled against hearing it. My father knew far more than I did from attending the trial for days after my testimony, so I would ask him questions. But then I couldn't bear to listen—my ears would start to ring, and my heartbeat would rev.

My father's ongoing presence at the trial marked a major turning point for our relationship. Growing up, I had always been close to my father—he was the more physically affectionate of my parents, quick with praise and a show of love. He'd also been my role model for how to live life to the fullest, and to never take no for an answer. He was the one I confided in, the one I thought of as my guide in work and life. My idol. But for years after the attack, I took a step back from him. On some irrational level, I think I blamed him for not being there to protect me. Even more so, I felt betrayed by his lack of emotion afterward. I think I expected my mother's numb response—I knew that's how she dealt with extreme emotional stress. But my father had never been withholding before. After the attack I saw in him no acknowledgment of sadness or fear—I saw only a desire to move on and to put the past behind us. This sent me the message that I wasn't allowed to be vulnerable with him—my daddy. In the years before the trial, I couldn't even hug him. He would put his arms around me, but I just couldn't embrace him. I would stand there like a board, bursting with grief and wanting to hug him back, but I could not wrap myself around someone who was closing me out. As a result, I was all alone with my shame and sadness—I'd not only lost my sense of self, but I'd lost my father.

During the trial, though, we reconnected. He'd call me in tears after a particularly difficult testimony. Now that he was vicariously living through what I had undergone three years before, it struck him how much he didn't know. When he finally showed me his own emotions, I broke down, heaving like my eight-year-old self used to do when he and my mom were leaving for a long trip. I had years of yearning built up in that cry, and I was finally able to tell him what his silence had done to me. He confessed that after the attack, when I was battered and bandaged, he'd only allowed himself to cry in the shower, because he was afraid that if he showed me how upset he was, it would be more painful for me. While I thought he was shutting down, he'd actually been trying to protect me. Just as nothing had ever prepared me for how to handle what had happened to me, he was fumbling his own way through an unfamiliar maze, not always making the right decisions about which way to turn, but doing his best. Now it all intellectually made sense—but I had felt emotionally abandoned for so long that it would take a long time for all the self-protective layers to melt away. Still, it was nice to be able to hug my father again.

After three years of fear and terror—three years in which I didn't know if the attack would ever be proved, and if I'd ever feel truly safe again—I received a call that the verdict had come back from the jurors.

My attacker was found guilty of attempted murder.

What I had never known, and was not allowed to hear during the first part of the trial, was that in the lineup, years before, Andrea and the men working on their motorcycles had all chosen number four.

I wasn't in the courtroom for the verdict, but it was important to me to be present when they sentenced my attacker. I wanted to see his punishment handed down and to know exactly how long he'd be put away. Would it be for five years, or nine years like his last sentence, with eligibility for parole? During the sentencing, I sat on the same side as him, but several rows behind, so that all I could see of him was the back of his head. I was so afraid of him still, even surrounded by people and court officers, that I actually wedged myself behind Jimmy on the bench.

The judge asked the attacker if he had anything to say before he was sentenced, and he said yes. Then he stood, and said that he had been falsely convicted, and that it was a case of mistaken identity.

It was like a gut punch on two levels. First—his voice. I had never heard him speak before. Throughout the attack, he hadn't said one word—not a curse or a grunt, a gasp or a sigh. He'd been as silent and singularly focused as a shark. It was sickening just remembering the utter soundlessness of his attack. Second—his words. All this time, I'd imagined him as psychotic and out of control, but there he stood calmly lying in front of the judge and everyone else.

I was dumbfounded. This man who I'd thought must be insane wasn't crazy at all. For a moment I was illogically terrified that the judge would believe him and that he'd somehow vacate the verdict. The attacker sat down, and we waited for the sentence.

The judge sentenced him to twenty-seven years *without parole* for attempted murder.

I don't recall a monumental sense of relief at the end of the trial, although that had to be one of the emotions that I was feeling. Instead it was as if the trial had roiled up so much inside of me—anger, anxiety, fear, abandonment—that I thought I'd explode from the pressure. After those initial waves of emotion subsided, I started to actually register the sense of closure elicited by the guilty verdict. The past was done, and I'd never have to go back there again. It was finally time to close this book, and write a new story.

After the sentencing, I flew back to the Bahamas—all by myself this time. I wanted to see with my own eyes that tiny island that my father had talked about. It turned out that there were several parcels of land for sale on the island, and so one day I hopped in a puddle jumper with a broker, and we flew over to see it. It was just as I'd imagined it—no houses, no running water, no infrastructure of any kind. It was peace and solitude, surrounded by turquoise water. At that point in my life I was a single girl sharing a rental with two other single girls. I didn't even own a car. But I'd been saving every penny I'd made for years, and now I knew why. On that trip to the Bahamas I spent every dollar of my savings on two parcels of land on a tiny island that didn't even have a dock. I never told anyone but my family, because I knew that most people couldn't possibly understand the sense in it. But it made sense to me. In my darkest moments—of which I already knew there would be many—I'd always have a piece of sunshine and warmth that was all mine. Even when I couldn't be there physically, I would keep its image firmly planted in my mind—pristine and beautiful.

PART III
Figure Eight

You must be the change you want to see in the world.

—GANDHI

CHAPTER EIGHT

Pins and Needles

Not long after my attack, my good friend Dianne gave me an article she'd pulled out of the *New York Times Magazine*. In it, a woman who'd been raped years before talked about the anniversary of her rape and how she marked it each year. I had always thought of anniversaries as happy occasions, but this was a time of mourning for this author, just as May 30 would forever become a day of mourning for me. She gave herself that time to sit with her pain, and then she put the mourning period behind her for another year and went back to the wonderful new life that she had carved out for herself. I aspired to that, and for years after I carried that creased and dog-eared article around with me, tucked into my wallet.

After the trial, I realized I couldn't keep spinning like a top, never risking a pause or a moment to look at myself and what had happened to me, afraid that if I stopped for one second I'd topple over. I knew in my heart it was unsustainable. This horrible mess inside me wasn't going to miraculously fix itself, and no matter how deeply I buried it, I couldn't pretend it away.

For the first time in three years, I was thinking about the future. I couldn't imagine what it looked like, but I knew I wanted it to be closer to what I had always wanted. I didn't want to just survive—I wanted to *live* again. Before I could reach that point, though, I had to work through—and relive—so much pain. I'd been digging a trench in the same circular pattern—running, running, running away from my past. I'd taken things as far as I could on my own, but this time I didn't have to go it alone. I called Ann and went back into therapy, not for the trial this time, but for me. The decision to admit that I needed help and to ask for it changed my life. Asking for help was something I had never been familiar with, so getting it made me feel like I was worthy of a future. And with that, I stepped off the track I'd been running on, and instead of making yet another left turn down the same old lonely path, this time I made a right. That right turn took me on a new path, and gave me hope that if I could do it once, maybe I could do it again. Perhaps I could view my life as an infinite number of figure eights—an endless pattern of opportunities, and new turns to take.

It was Ann who helped me realize that it was useless to ask, "Why me?"; no answer would ever be good enough to justify what I went through. I spent hours in her office in tears, on some level hoping that she'd just tell me what to do, and then I'd be healed. But there was no magic set of instructions, and the first step was the hardest. Part of my new pattern was acknowledging that my survival mechanisms of the past weren't working anymore. Putting that box far inside me back in Boston had served me for a time, but that time was over.

Facing my darkest fears—not just about the attack itself, but about my very existence—enabled me to distance myself from the attacker once and for all. During the lineup and the trial I had been terrified of his physical presence, but now that he was behind bars for a good long time, I made a conscious decision to separate his actions from me and to take back the power he'd had over me. It's not as if I was suddenly healed, but increasingly I felt that there was a barrier between my present life and the past.

I was beginning to believe that I wasn't at fault for what had happened to me. For so long I had thought that I had been chosen because I deserved it in some way. I still had so far to go in overcoming my deep shame, and doubts about my worthiness. But I hoped that someday I might be like that woman in the article, eventually able to memorialize the pain on just one day, then put it behind me until the next anniversary.

Back in college, when Nicolette, Andrea, and I were living together, our apartment was broken into. Andrea had bragged afterward that if she'd been there when we were robbed, she would have physically attacked the thief. One night when a drunken woman came barreling past us, Andrea started screaming after her, and insisted that no one could push us around. Of course I knew that those were just fantasies, and I meant what I'd told her—that none of us can know what we'd do in a situation like the one I'd endured.

Andrea and I stayed friends through the years, even after the trial, but it would never be the same. One day I was with Andrea and her family, and her mother said something to me like, "Thank God Andrea was there, she saved your life by opening up that door." I felt so devastated in that moment. Andrea hadn't saved my life—*I* had saved my life. My eyes stung with the effort it took to hold back my tears and clamp shut my mouth.

Our relationship waxed and waned. I had continued to defend Andrea's actions over the years, and it never seemed to me that she ever appreciated it, or valued my loyalty; in fact I often felt *she* resented *me*. Often, we'd make plans, and she simply wouldn't show up. I'd go to her neighborhood to make it easier for her, and she'd leave me sitting in a restaurant or standing outside a movie theater waiting . . . and waiting. I felt like a blown-off blind date. When I'd call her to find out what was keeping her, she'd say, "Oh, Gilbs, I'm sorry. I spaced." Each time she stood me up, I would get upset all over again. Eventually I had to confront her, to give her the chance to explain. I thought, This is my close friend—she would never hurt me on purpose. Maybe I'd done something to offend her, I reasoned. And of course she couldn't read my mind, so if

I didn't say anything to her directly, then I'd just be holding a grudge.

I'm all about the conflict resolution, not the conflict. So the next time I saw Andrea, I said, "I don't know why you keep doing this to me. It really hurts my feelings." I'll never forget how she looked at me and said in reply, "You know what, Jen? I think you expect too much from people. It's too hard to live up to it." And that's when it finally hit me. I'd never expected anything from her, not even to try and save me. But it was *her* guilt and frustration at who she really was on that May day that was coming between us, and I think it was too much for her to be around me.

Andrea and I had been as close as sisters before the attack. She'd been my roommate for three years in college, and she was the one person who knew firsthand what I'd experienced. There were moments when my memories of the attack were so alien and horrifying to me that it was a comfort simply to have a witness who could attest that what had happened to me was real, and not a bad dream. It anchored me to know that she was there, and I fought hard to continue our connection. But our relationship was pulling her down to a place she didn't want to go.

All the emotional work that I did after the trial was exhausting, but once I got past the very worst of it, something unexpected began to happen to me. Without being fully aware of it, I started loosening the tight grip I'd been keeping on my emotions. Slowly but surely, I felt the feeling flow back into my numb heart.

I truly had no idea how tightly wound I was—and to what degree I'd made myself impermeable—until I started to release my hold. Ironically, at a time when I most needed to feel joy, I'd actually cut myself off from one of my simplest pleasures in life, and without even realizing it.

My mother raised us all to appreciate flowers. She was a master gardener and a docent for the Brooklyn Botanic Garden, and she wrote a gardening column for the local paper. She so loved her garden that she hated the very idea of edible flowers—she said it was like cannibalism. When I was little, she'd take each of us girls on tours of her garden, and by the age of five I could name every single flower. Even more so than the color of all those flowers, it was the scent that I loved. I remember being enveloped by their fragrance, just completely drunk with it, even as a child. To this day, flowers and their aromas bring me home to my mother's garden.

As an event planner, I could never bear to see how many gorgeous flowers would be dumped in the trash at the end of the night. Catering staff will often take home the food that's worth saving, but the flowers were just tossed away. I generally passed on taking home the sweets, but I could never bear to see all the gorgeous floral arrangements treated as garbage. So while the waitstaff was packing up boxes of cookies, I was stuffing big plastic bags with flowers—enough to fill a taxi. At home that night I'd stick them in every vase and bowl we owned. When Rachel woke up to the floral explosion in our apartment the next morning, she'd look around and say, "Oh my God, who threw up roses in here?"

I loved going through the flower district with clients. I soaked up all the colors, and I could still identify all the flowers. In the first few years after the attack, though, one key aspect of the experience was missing for me, and I wasn't even aware of it until I made a discovery one spring day.

I was walking to my office on Fifty-eighth Street between Eleventh and Twelfth Avenues. It was a warm day in May, exactly one year after the trial, and four years after the attack. May had been a dark month for me in the last four years, but this year was different in a way that I don't think I registered consciously. I recall a sense of lightness in my heart. I remember enjoying the sensation of the warm sun on my skin, and feeling my old excitement that the days were getting longer and summer was coming.

Now that whole block is condominiums, but back then it was abandoned buildings and empty lots covered in broken glass and enclosed by wire fences. There was nothing alive in that mass of concrete for blocks. I was walking along, noticing how the bright sun made the shards of glass glitter, and suddenly I smelled something—a waft of sweetness. I stopped dead in my tracks and inhaled the scent deeply into my lungs. It was honeysuckle. It reminded me of my childhood—of my mother's garden and summers at camp. Then I saw a honeysuckle bush peeking out between two ruined brick walls. It brought tears to my eyes, the smell was so sweet. Then I cried real tears. Oh my God, I realized in that instant, I had lost my sense of smell, and it was back for the first time in four years. I stood on that street and wept. The things I had denied myself without even knowing it. My sight and hearing had betrayed me on the day I was followed—I'd never seen or heard the attacker behind me, all that way from the subway. Ever since, I had become acutely attuned to visual and auditory signals. I always knew exactly who was on the street behind me, or near me on the subway platform, or running behind me in the park. In the process, my senses had become so tightly controlled that somehow my ability to smell had atrophied—neglected and unused, it just withered. But now it was coming back to life.

Smelling that honeysuckle was a moment of innocent joy. It was reassurance that there was something pure inside me that couldn't be forced, or restrained, or manipulated. It was spontaneous, uncontrollable, and completely genuine. It gave me hope. There would always be a before and after in my life, but on that day I began to believe that in the "after" I would be okay.

Not long after that day, I was hanging out with Laura, and she said something that was really funny to me, who knows why. Pretty soon I was just rolling with laughter—the kind where you laugh so hard it makes you laugh even more, until you forget what you were laughing about in the first place. I laughed so hard that I snorted—and there it was, my all-out, totally unladylike guffaw that I'd also left far behind. I laughed until I cried, and then I cried for real because I realized that I hadn't laughed with that kind of reckless abandon in years. Until that moment I had always held back just a little bit—I kind of stood to the side and observed other people's mirth from a safe distance. Maybe I'd laugh to acknowledge that something was funny, but it wasn't *my* laugh.

After the attack, parts of me had fallen into a sleep state, while other parts of me took over the operations. My giddiness and vulnerability had been sacrificed for the sake of staying on guard and protecting myself. After the trial, though, and thanks to my work with Ann, I started to feel a little shift. First my laugh came back, and then there was a palpable little tingle in some of those other sleeping parts of me. I wanted to regain that sense of adventure that had been such a strong part of who I was before.

Jimmy had formed a protective cocoon around me for more than two years. I'd been as happy as I could be during that time, and I'd stretched my battered little wings within the boundaries of our safe life in New York. During the trial Jimmy was truly my rock, and I had held on tight. He was safe, warm, and wonderful. But through no fault of Jimmy's, that safeness started to feel constricting to me.

Just prior to the attack, I had moved home after traveling around the world on a boat, living in Europe, hitchhiking all over Sicily, navigating through Russia and Turkey. After the attack, the farthest I'd traveled was the Florida Keys. Jimmy felt like my family, but I was antsy and impatient, and the girl I'd been for the last few years no longer felt like the real me. I didn't know exactly who the real me was anymore, but I wanted to find out. And until I did, I couldn't really give myself to anyone. Timing was just not on our side.

When I broke up with Jimmy for good, he cut off all contact with me. I was devastated, deeply conflicted, and genuinely fearful to lose him so completely. But at the same time I felt a deep desire to be on my own, and to see what the rest of my life might hold for me.

Over the next few years, pieces of me that I had carefully compartmentalized began to flow back together. I began to feel less splintered, less broken—and less afraid. This made me more confident in my work, too. When every event stopped feeling like my own personal Waterloo, I realized that solving problems wasn't a live-or-die necessity or an adrenaline-pumping thrill, it could actually be a genuine joy.

For a full year I had been planning a massive alumni event for a graduate school in Manhattan. Fifteen hundred people were expected for a seated dinner dance that we'd hold on Piers 60 and 61 on the Hudson. It was really like planning three separate events, because each class celebrating its five-, ten-, or fifteen-year anniversary was given its own dedicated space. At 4:30 p.m. the day before the event, I received a hysterical phone call from the alumni director. She was at the opening-night cocktail party, and she was being screamed at by alumni who'd been told there was no room for them at the dinner dance. It turned out there were 350 extra alumni who'd flown in from all over the world, just assuming they could buy their tickets for the dinner upon arrival. So now it was no longer a seated dinner for 1,500 people—it needed to be a seated dinner for 1,850 people. But the dinner was on the pier, and the space was filled to capacity. The alumni director was hyperventilating, completely unable to imagine a scenario that didn't involve losing her job. I said, "Okay, give me twenty minutes."

I hung up the phone, and I didn't have the foggiest idea how we were going to swing this. Could we tent the parking lot? Was I really going to put a formal dinner on uneven asphalt? Then it hit me. It was such a perfect solution that I actually burst out laughing. I couldn't build another pier, but I had a better idea. I called the owner of a party boat that I often booked and asked him (a) if he was free the following night, and (b) if he could get a permit to dock at Pier 60 in the next twenty-four hours. The answer to both questions was yes, and I was able to call my client back and save her weekend and possibly her job as well.

I had proven myself worthy again, but this time I'd done it minus the fear. It wasn't my fault that 350 more people had shown up, and I didn't believe I'd never work in this town again if I didn't solve the problem. But I'd solved it anyway—and I'd done it with ingenuity, and creativity, and yes—joy.

I began to think about taking a vacation, something I hadn't done in years. In the past, I'd been too afraid that my business would self-destruct in my absence. I had always felt that I had to do everything and be everywhere at once. I'd be out at a club until 4:30 in the morning, and then be back at my desk at 7:30 that morning. All that effort to be fabulous and perfect was exhausting, and once a year or so I needed to step off the merry-go-round and be quiet and still and (for the most part) by myself.

Even though I hadn't been anyplace exotic since before the attack, I kept folders in my desk of everywhere I longed to see one day. I ripped up my *Travel and Leisure* and *Departures* magazines, collecting names of restaurants and hotels the way cooks collect recipes. I had been to Israel on a summer program in high school, and had wonderful memories of it, always wanting to return. My other obsession was Italy—I had been there during my stay in London, but had never been to the Amalfi Coast. So when I did start to travel again, it was only natural that I returned to Israel and Capri.

When I landed in Israel, usually at two or three in the morning, I'd scope out someone young and single in baggage claim who looked like a local. Then I'd walk right up to him and ask for help getting my bags to a taxi and negotiating the price—since I knew I'd end up paying less if the person doing the talking was male and spoke Hebrew. On one particularly great trip to Israel, I met my perfect tour guide on the layover in London. I'd traded in my miles for a first-class seat, and the whole way to Tel Aviv, Inon came up from coach to chat with me. He was a banker living in Brussels, but formerly he'd been an expert in martial arts for the Israeli army. He was at least six-three and good-looking, and how could I refuse when he asked for the number where I'd be staying? A handsome Jewish banker who knew how to fight wanted my phone number—I could barely suppress a schoolgirl giggle.

Inon put me in a taxi when we got to Tel Aviv, and by the time I got to my friend's sister's house, it was 4:30 a.m. At 8:30 that morning I was drinking coffee in the kitchen with Aya when the phone rang. She spoke Hebrew into the phone and then smiled and handed me the receiver. "It's for you," she said. It was Inon. He told me that he was coming to Jerusalem, and that he wanted to take me around Israel. I didn't know him at all, but I didn't even hesitate.

I knew what it was like to be in a genuinely life-threatening situation, and maybe that gave me a special kind of radar. But I never once felt unsafe or alone when I was traveling by myself. Especially in Israel and Italy, I felt at home, among kindred spirits. The Israelis seemed to know what it was like to live precariously and just enjoy the moment because there might not be a tomorrow. They didn't waste time; they never procrastinated when it came to fun or a new experience. There was no game playing. If you met someone and he liked you, he told you so. If he wanted to see you, he didn't make a date for next week. He said, "Let's go—now." Their answer to any new experience wasn't *Why*, it was *Why not?* You have one life. So make it count. *Yes* felt like the right response to me.

The Italians also seemed instinctively to know how to enjoy the moment, and I loved how completely anonymous I felt when I was there. They appreciated beauty and food, and could revel in a simple sunset in a way that I'd never experienced in the States. It was just the dose of freedom and immediacy that I needed. Back home, I was always trying so hard to be the person that I thought I should be. There was something about removing myself from home and being completely unknown in a new place that made me feel whole again. I could be quiet and alone when I wanted, or I could go to a restaurant and strike up a conversation with some locals. No one judged me, and there were no explanations needed.

I think that one of the reasons I loved traveling alone is that it gave me an opportunity to celebrate the self-reliance that I'd embraced (for better and worse) after the attack. I had learned that no one else could magically make me all better or make the bad thoughts go away. I needed to redefine what it meant to be alone and make that concept feel strong, powerful, and satisfying. It was very important to me to know that I had the resources to fend for myself.

As risk-taking as I could be, I wasn't crazy. I definitely had rules. I didn't travel to little towns by myself, and I never got in a car with a stranger. One of the reasons I loved Capri is that once I was there, I could reach everything on foot. I'd stay up late and sleep late, and I'd spend the day alone on the beach reading and taking long walks, enjoying the peace. If I felt like company, I'd have a drink at a bar. Once I was befriended by the son of the owner of the modest little pensione where I was staying, and he and his friends invited me to dinner and dancing. On nights when I started eating dinner by myself, just me and my book alone in a restaurant, it rarely lasted long. Inevitably I'd be invited to join someone else's table, and then asked to a party on someone's boat.

I didn't always travel alone—I also took some epic trips with Deanna, and on those trips I would be in full-on fabulous mode. On those giddy, wonderful vacations I felt like I was reclaiming my lost years. We were young and broke and stayed in nasty pensiones, but that didn't matter because we hardly spent any time in our rooms anyway. In Paris, we stayed with friends of friends on the outskirts of town. It was roughly the equivalent of visiting Manhattan but staying at LaGuardia Airport. We got all tarted up in tiny dresses and red lipstick and rode the metro at least an hour into the heart of Paris. Nicolette had given me a list of the hottest places to go for coffee, dinner, drinks, and dancing. Back then *the* nightclub was Le Bain Douche, and by the time we arrived, the line to get in was hundreds deep. I decided that if I didn't wait on line in Manhattan, then I was not going to wait on line in Paris, so I grabbed Deanna and we trotted up to the front of the line. We were hand in hand in our sparkly party dresses, and the lesbian bouncer waved us right inside.

It was already midnight, but the tiny disco was bizarrely empty and quiet. We got ourselves drinks at the bar, and soon Deanna was talking to one man, and I was chatting with another. At the exact same moment each man said, "So, are you here for the Prince concert?" Each of us nodded, *mais oui*, and then mumbled *excusez-moi*. Then we found each other and silently squealed our disbelief. We were in Paris, in an empty nightclub, and in a few hours Prince was going to give a private concert. This could *not* be happening.

The club filled up, and Prince came on around three in the morning, in a space no bigger than my living room. I still remember the exhilarating feeling of walking out onto the streets of Paris at seven, sweaty from dancing all night, to the sound of street sweepers and the smell of croissants just coming out of *boulangerie* ovens. It was magical.

Times like that, I didn't feel like I was trying quite so hard. My joy was spontaneous and real, and there was nothing forced about it. The feeling wasn't permanent—not by a long shot—but I embraced it body and soul. I was opening up to a colorful world full of possibility. I'd be back to reality soon enough, but until then, I seized the moment.

CHAPTER NINE

Bicycle Pants

For many years after the attack I was convinced that everything in life was a bargain—to get one thing, you had to give up something else. The Jenny who'd walked into Andrea's apartment building thought that she could have it all. The Jennifer who came out the other side of that horrible experience knew that nothing came without a price.

I chose the perfect line of work to prove that theory. So often, no matter how hard I worked, something awful was bound to go wrong, and I'd have to pick up the pieces. Being hypervigilant in my line of work wasn't paranoia—it was necessity. If anything was going to work out well, I reasoned, then it was because I'd worked my butt off, given twice my all, and still expected the absolute worst. Nothing could be taken for granted, because the second I got comfortable it could all come crashing down.

Once we took on an important event for a private law firm headquartered in Europe. If events are disaster magnets, then planning an event on another continent is a disaster invitation. The margin of error becomes canyon-size when you factor in time changes and jet lag and customs and language barriers. The conference for two hundred guests was planned for a distinguished old hotel in Portugal. Two months prior to the event, the organizers decided to double the number of guests. However, this antique hotel didn't have a banquet room large enough for twice the number, and it wasn't possible to move the event elsewhere. The only solution was to build a temporary structure on the grounds of the hotel. Then, with just a few days to spare before the event, the local tent company informed us that they didn't have the correct materials available. Luckily, private European law firms at that point in time didn't lack for cash—this was a company that could afford to bring in the most famous opera star in the world as their entertainment—so when ingenuity failed us, we were able to throw some money at the problem. We paid triple the usual asking price for a company in Spain to truck the tent to us overnight, and to construct it in round-the-clock shifts.

The temporary structure was gorgeous, and that evening, with just an hour left in the program, my staff began to silently celebrate. I, however, never celebrate until the last guest has headed home. Of course, with just forty-five minutes left, a storm descended in our area, and my staff and I, along with the triply paid Spanish construction workers, noticed that the wind had actually caused the back wall of the structure to buckle. We all bolted for the door—and in a wind that would have sent Dorothy flying to Oz, we physically held up that wall until the event was over and all the guests were tying one on in the hotel bar. It was truly insane, but that was my life. I never, ever took anything for granted.

It was the same thing in my personal life. I was at a point where things were looking up in so many ways—I'd put the trial behind me, my business was growing, and every day I was allowing myself to enjoy life a little more. But still I could never let myself believe that I deserved any of it, or that it couldn't all be taken away if I let my guard down for even a second. It was two steps forward and one step back. So, to push the odds in my favor, I started to make endless deals.

I'd been obsessively running ever since I'd moved back to New York, but it wasn't until sometime after the trial that my calorie vigilance became fanatical. When I lost weight during those anxiety-filled weeks before the hearing, I discovered two very powerful realities. One, I liked my new thin, lithe body; and two, while I could not control the outside world, I could control everything I put in my mouth.

It was around this time that my relationship with food became complicated. Gone was the girl who'd made drunken trips to the campus baker and stuffed herself with cookies and brownies. I found a new outlet for all my inner fears and brokering: making food sacrifices. For several years after that, I cut out whole food groups, one at a time, carbs among them, until I regularly ate nothing but egg whites and turkey breast. Here I was an event planner, constantly surrounded by food—the most exquisite cakes and luscious hors d'oeuvres, homemade breads and pastas—and I walked by all of it as if it wasn't there. I'd plan the most gorgeous wedding, I'd cry during the ceremony, I'd be just as giddy as the bride during all the festivities. But I wouldn't allow myself a bite of the wedding cake at the reception. And it's not as if I was the kind of person who'd never been much interested in sweets. I *loved* sugar. I could have happily lived on nothing but brownies and alcohol. But I denied all my natural cravings so successfully that eating only lean protein and undressed greens became a way of life.

I now closely regulated everything that I ate. The experience of eating became another arena in which to prove myself. It was my next deal with the universe. *If I deny myself this*, my unconscious reasoning went, *then maybe I'll be able to keep these other things that I fought so hard for.* It seems so contrary to logic—why would a person who had been through something so awful want to punish herself further? But anyone who's been afraid to speak a wish aloud or uttered a *kinehora* (Yiddish for "knock on wood") knows the fear of losing it all if you dare to get complacent. For the same reason, sacrificing and denying myself pleasure became my ways of inoculating myself against further disaster. The constant vigilance was exhausting, but I figured it was the price I had to pay for being alive, and being safe.

My secret insurance policy was certainly a form of self-punishment. But there was also something about the knowledge that I could withhold food from myself—that I could abstain and resist—that gave me a feeling of power. I remember actually liking the feeling of hunger, and knowing that I could go without. When I'd go out to restaurants with friends or on dates, I didn't even look at the menu. It was a minefield of things I wouldn't allow myself—pasta, bread, butter, dessert. I'd order broiled fish, a steamed vegetable, and a salad with vinegar (no oil). Some people probably thought I was nuts. Others admired my discipline. For me it was just mind over matter. Not eating when I was hungry was like sprinting that extra mile. I did it even when it hurt. I did it *especially* when it hurt. I did it because I could.

My body became a battleground in more ways than one. In addition to my exercise and food denial, I was also still deeply self-conscious about my scars. For years after the attack I refused to wear a bathing suit in public. The beach on a summer day became a terrifying gauntlet of exposure. Initially, I told myself that I had good reason to cover up at the beach—the scars would heal better and be less noticeable in time if I protected them from the sun. But even once the scars had faded to barely noticeable white lines on my thighs and butt, I couldn't bear to reveal myself.

Black, knee-length bicycle pants became my second skin and my perfect excuse. In the Hamptons, when everyone else in my summer-house was headed to the beach, I'd head out running. Later, I'd meet my housemates on the beach on the way back from my run—still wearing the bike pants. I felt such potent shame and anxiety, and inside I died a little each time someone said, "What's with the bike pants, Jen?" Or, "God, you're making me hot just looking at you." Or, "C'mon in the water with us, Jen."

I wanted to, but I just couldn't. Those pants were my emotional girdle—they held me in—and I just wasn't ready to lose them.

I'd grown up fearlessly swimming in open water off my father's sail-boat, and I'd always adored the ocean. But once I started covering up, it became a compulsion to protect my secret. Here I was renting a share in a beach house, while the whole topic of the beach was traumatizing for me. I'd find excuses not to hang out with my housemates all day long because it gave me such angst.

None of those people in my Hamptons share even knew I'd been attacked, and I'd always refused to be that girl who exposed her tragedy to everyone. When my housemates asked me about the bike pants, I'd say that I wasn't hot, or that I was about to take a walk or go for a run. After a while the girls mostly stopped asking me about the pants. They realized that whatever my issues were, asking me questions about it wasn't helping. The men never stopped asking, though. I'm sure they all thought I was just being sensitive about my body (I remember one of them saying, "What, do you think you're fat or something?")—and of course I was sensitive, but not for the reasons they might have expected.

The bicycle pants imprisoned me as much as they protected me. I had the craziest tan lines from wearing them all the time. I'd wear a bikini underneath, so the top of me was bronzed, and then there was a white band of flesh that started at my waist and extended all the way to my knees. This meant that once I committed to the bicycle pants for a summer, I was stuck. If I took them off, I'd look like I just emerged from a body cast. I couldn't even wear a short skirt for fear of exposing that weird line of demarcation.

One horrendously hot day I went for a walk by myself over to the family beaches, where I didn't know anyone. I walked as far as I could, dripping with sweat and looking longingly at the ocean. Women and men of all shapes and sizes were in the water, walking on the sand, playing with their kids, letting it all hang out. And here I was, young and thin and half encased in spandex.

I looked around me to make sure that I didn't know anyone, and I just thought, Screw it. I peeled off the bike pants, and I was fluorescent white underneath. I went into the ocean for the first time in years. I'm sure I cried from the shame and sadness. The whole charade was just so pathetic—and exhausting. I so badly wanted to be a normal person. I was surrounded by people who weren't supermodels by any stretch, but they all seemed comfortable in their skin—or at least comfortable enough to wear a bathing suit in full view and jump in the water on the hottest day of the year. Why was I such a loser? Such a freak?

That internal monologue of hate speech was familiar to me. Whenever I'd inevitably fall off my dieting wagon and binge on candy, I'd wake up the next morning with a headache and a sick feeling in my stomach. Then I'd be positively evil to myself. I'd tell myself that I was weak, gross, disgusting. For the next several days I would physically punish myself. I'd drink liters of water but wouldn't eat until after 4:00 p.m., at which point I'd allow myself a head of lettuce. Then I'd go for an eight-mile run. It wasn't just that I had a crazy notion that eating a bowl of ice cream would leave a literal, immediate imprint on my butt. That was insane enough. But my grip on my weight was about even more than that. One night of eating chocolate had the power to negate years of walking my tightrope of deal-making. Just one brief loss of discipline terrified me.

That year in the Hamptons, I met someone who seemed (superficially, anyway) so unlike me that we never should have become friends. Bennett was one of the housemates in the new share I'd taken. The first time we ate in a restaurant together, he blew my mind by ordering chocolate soufflé as his first course. He said, "Why not eat the best part of the meal first?" He ordered steak with gorgonzola after that. I ate my dry salad and fish while experiencing total food envy. I kept myself on such a tight leash, and here was a man who had no leash. That was such an alien notion to me—I thought you had to work for everything in order to deserve it. He was the opposite—he didn't question whether he deserved life's pleasures. For me, running was far more about control than release; Bennett played racquetball because he enjoyed it, and that's how he lived every day.

Bennett completely intrigued me the first time we met. He was eight years older and seemed so secure—in his skin, his career, his family and circle of friends. He could be reserved and a little quiet at first—he liked to take things in and let other people do the talking. The son of a New York State Supreme Court judge, Bennett grew up in a highly intellectual family where the regular practice at dinner was to conduct a debate over a chosen policy subject. He was absolutely brilliant and destined to be a lawyer, but after his father died when he was in college—a memory that still haunted him, although he'd never let on as much—he took a different turn and ended up a successful executive at Bear Stearns.

Bennett was stylish and sophisticated, with dark hair and eyes and skin that tanned deep brown in the sun. He always seemed to have his own plans, and he'd leave the rest of us in the middle of the day, all dressed up in a linen suit, and zip off in his convertible to a polo match, or to some fancy party on a yacht. He was a wild combination of stoic on the surface—you'd never know what he was thinking unless he told you—and charismatic, and given to the most grand and beautiful gestures. He cultivated orchids, owned about nine hundred bow ties, and was the only straight man I'd ever met who could confidently pair a pink polka-dot tie with a purple-striped shirt. And he *only* wrote in green ink.

Bennett loved type A women, and he quickly adopted me. In the city we'd often have dinner. I'd talk a mile a minute, filling in any silence, but he never felt pressure to make small talk. Occasionally he'd call on the spur of the moment to see what I was up to. Once he called to see if I was busy and I told him I was shopping for a new suit, so he offered to come along. I tried something on and went out to see what he thought. He looked at me carefully and said, "You can do better." He was right, and I was stunned. I'd never had a man be so honest with me about my appearance, and I could not get that suit off fast enough.

I knew Bennett had another close friend named Jessica, and I was always curious about the attention he paid to her. Often he'd leave the Hamptons on Saturday afternoon, saying that he had promised to see a play with her. I thought, Okay, a friend doesn't leave his share on a summer Saturday just to go hang out with another friend. So finally I asked him if she was his girlfriend. He said no. He was obsessed with her, but they were just friends. I didn't blame Jessica for not wanting to lose Bennett as a friend, even though she wasn't interested in him as a boyfriend. When he focused his energy on you, it was like a spotlight, and it was a wonderful place to be. And I found that I could live vicariously by making him happy. I knew he loved chocolate, so I started bringing him special treats from my events. Every time, he ate it right there in front of me, and knowingly gave me the satisfaction of his pleasure. He knew I was insane about my own eating rules, but never judged me or ridiculed me. He just let me be. He was that familiar warm hug I always sought out. We became the absolute best of friends.

Bennett had a direct hand in forcing the issue of the bike pants with me. I didn't shed them right away after becoming friends with him, but he managed to make the first dent in the armor. He had a friend named Mark who had a very dry, sarcastic sense of humor, and weekend after weekend the two of them would harass me about the bicycle shorts. One day Mark said, "Here's Jen in her bike shorts again. Are you going to be buried in them?"

That was it, I'd had it. At an opportune moment I took the two of them aside, and I said, "Okay, you want to know why I wear bicycle shorts? Here you go." I told them just the facts, and I didn't get emotional. I said, "X years ago I was stabbed Y times with a screwdriver, and most of the wounds are on my legs. For years I didn't want them to scar, but now covering them has become habit as well as self-consciousness." I told them that whenever anyone asked me about the scars, it was an unwelcome reminder of what happened to me, and I just preferred not to share that part of me. When I'd finished, I asked them both to keep it to themselves. They were shocked and quiet, and they promised they would never speak of it. They were both true to their word.

I'd never told anyone other than a boyfriend about what had happened to me—and even then it was only out of necessity. Now the circle of people who knew the real me had gotten a little bit wider. I was starting to test the waters of honesty about my past—gauging their reactions and my own comfort level. Bennett's complete acceptance of both sides of me—the vulnerable and the tough—was a revelation, and it was the start of a new phase in my life.

. . . .

To me, Bennett was like a brother. But to other women he was catnip, and he had a series of lovely girlfriends. One year it was Stacey, and they were obviously very serious. They were always cuddling on the couch and would make their own plans most nights. Once Bennett and Stacey were having a predinner drink at the house, and he asked me if I wanted to come to dinner with them. I had no plans, but the last thing I wanted was to be the third wheel on Bennett's date, so I put on my trusty bike pants and headed out for a run.

The driveway leading down from the house was steep, and I was gone not two minutes when I wiped out on the pavement and came limping back up in tears. My hands were bleeding, my knee was bleeding, and Bennett was all over me with ice and bandages. He refused to go to dinner without me, so I got cleaned up and went along with them, hands wrapped in gauze and all.

That summer Bennett and I were both paired up with other people—he with Stacey, and I had met Rich, a man who looked good on paper. He was handsome and successful, he came from a wonderful family, and we had a lot in common; he seemed to be everything I thought I wanted. Except we fought a lot, and I felt like he was competing with me all the time.

One weekend in the Hamptons, Bennett and his girlfriend and Rich and I went out to a restaurant together. I excused myself to go to the bathroom, and on my way there I ran right into Jimmy, whom I hadn't seen or spoken to in two years. My heart flew into my mouth, and I could barely speak. He seemed just as flustered, and quickly explained that he was there to pick up a pint of ice cream for him and his fiancée. *His fiancée.* I'm not sure which exploded first, my heart or my brain. We said our good-byes, and I staggered into the bathroom to bawl my eyes out.

Seeing Jimmy happy and engaged to be married jolted me into realizing the degree to which I wasn't where I wanted to be in my life. Maybe we weren't the right people for each other, but still I was shocked at how emotional I felt to see him marching on with his life while I was still spinning, and running, and seeking. I was planning everyone else's adult milestones, and meanwhile all of mine were still out there in my uncertain future. Jimmy had moved on, and here I was with a man that I didn't really want to be with—that, truth be told, I didn't even like all that much. I thought, What am I doing?

I don't remember much about dinner after that. When we got back to the house that evening, Rich and I got into an enormous fight. I was done—I was tired of forcing the relationship, and I wanted out.

It was a long, cold night in the Hamptons after that, and on our way back to Manhattan the next day, Rich continued to berate me. He couldn't grasp why I was such a wreck over an old boyfriend. He was so confident in his own status as a brilliant catch that he simply wouldn't accept that I wasn't happy with him. It wasn't that he loved me—I knew he didn't—it was that his competitive juices were flowing. Finally, exhausted by the screaming, I told him to pull over, and I got out of the car.

I figured Rich would drive around the block and come back for me after we'd both cooled off a bit, but no. Rich never came back, and I was stranded on the side of Montauk Highway. I only made one call. I called Bennett. He was just about to head back to Manhattan with his girlfriend. And of course he answered my call. I blubbered one long stream of gibberish until they finally figured out my general coordinates. Then he picked me up, without saying a word. That was the thing about Bennett, he just showed up for me, every time. And that meant everything.

CHAPTER TEN

The Scary Mask

My work was my life during those years. It was all-consuming, and that's just the way I wanted it. My goals were to make the clients happy, get more and bigger clients, and make my company a success. This was very tangible, and easy to measure. Business was a daily challenge for me; someone was going to win a job, and someone was going to lose, and who likes to lose? That took all of my emotional energy and focus. And while I was certainly more secure in my abilities than I'd ever been before, still I wouldn't allow myself to relax long enough to examine what exactly I was building, and why.

I had six women working for me—all as young or younger than I was—and I had absolutely no idea how to be a great manager to them. My only concept of that relationship was that I needed to be in charge—a benevolent dictatorship was probably an accurate description. I couldn't be friendly and also expect my staff to respect me, so I put on a mask of all-business.

When I walked into my office in the morning, a tense silence fell over the room. I never said good morning, or asked anyone about their weekend. I wasn't being unfriendly on purpose, but as I walked through the door I'd already be in the middle of a thought about who I had to call or what event we had that week, and I'd fire off follow-up questions from the day before. If the office got too chatty, I'd look up from my desk, and everyone closed their mouths and went back to their jobs. I'd forget to say, "Good work," or "Nicely done." It never occurred to me that people needed to be encouraged. I wasn't motivated by praise—if I did a good job, we'd get more clients.

I liked to think that my staff knew I was on their side, and realized that if they ever needed me, I was there for them. If a client ever treated them rudely, then that client was either fired or put on notice. But when it came to the kind of casual office exchanges that most people didn't even have to think about, I was hopeless. I remember someone actually telling me they heard that I made Anna Wintour look like a pussycat, and I actually laughed. I thought it was a compliment, sort of. Looking back on it now, I think I was terrified that if I showed one tiny hint of vulnerability, the floodgates would open and I wouldn't be able to keep everything in check.

There were still so many days that I woke up after nightmares and bad sleep. I'd walk into the office with a leaden cloud over my head. Later, after I'd scared the bejesus out of everyone, I would sit in my office and think to myself, Why can't I be normal? It hurt my feelings when the women in my office would head out to drinks after work without a thought of inviting me. I didn't want to be the way I was, but I couldn't help myself. And I didn't blame my employees for not realizing there was a human being under all my armor. I'm sure they would have been shocked to know what an insecure mess I was inside.

The only upside to this alternate persona I put on for the world is that it seemed to give me insight into my clients and their own private moments when they thought no one else was watching. Didn't everyone wear some sort of mask? There was the A-list socialite who secretly worried that no one would come to her party. One stunningly beautiful bride seemed so sure of herself but lost it when she was faced with performing her first dance; she never liked to be the center of attention. Inside, I found out, she was just as self-conscious as anyone. One particularly shy bride, a "plain Jane" type, showed a totally different side of herself when she went for the vampiest, skintight wedding dress. She was the youngest of five kids, and nothing ever fit quite right during all those years of hand-me-downs. She didn't want to look like anyone else in her family on her wedding day. From all these women I learned that I wasn't the only person in the world with something to prove. It occurred to me I could never judge anyone again. I had to have an open heart and mind when I met people. Maybe the outgoing woman I was speaking to had her own secret story. If anyone judged me based on first sight or my tough exterior, they would have gotten it completely wrong.

This was a profound insight for me. I'd always believed that what made me truly good at my job was my perfectionism. Now I was discovering that my far greater skill was that I could sense the hidden aspirations that drove my clients to want to write a fresh new story on their special days. My mission wasn't to be perfect. It was to surround myself with people who were celebrating, and to know that I had helped them make their joy tangible. I hadn't yet figured out how that could be better reflected in my office itself, but it was a first step.

When I first got into the business and was chasing after corporate clients, it was all about being fabulous and impressing the clients with how hot, cutting-edge, and stylish I could make their events. But now I was evolving in my sense of what was really important, and slowly I began to focus much more on the emotional content of my events, the meaning behind the celebrations. And while I immersed myself in my clients' emotional milestones, something inside me was clicking into place as well.

One client in particular brought home for me how much I had changed since my attack. Michael was a highly successful Internet entrepreneur at the very height of the boom. He was good-looking, wore $3,000 handmade suits, a gold Rolex, and custom-fitted shoes, and had a cultivated swagger, but inside lurked the insecure science geek he'd been in high school. He was head over heels in love with his drop-dead-gorgeous girlfriend Alana—a girl who wouldn't even have glanced his way ten years before—and wanted to throw a surprise party for her thirtieth birthday. He gave me carte blanche to make it memorable—he actually told me that he wanted me to design a night that would impress even me. He'd been to a million standard-issue over-the-top parties, and he wanted this night to be *unique*. What touched me more than anything was that he wasn't just some rich businessman with a trophy girlfriend. He *loved* her. And he felt like the luckiest man on the face of the planet that she loved him back.

I don't think I have ever had a more nervous client. He was anxious about everything. He was desperate to make his girlfriend happy, and afraid no one would come. And if people did come, he was worried that the evening wouldn't be fun—then all his guests would find out that he really was just a big nerd. I calmed his nerves and assured him the event would be everything he wanted and more, and then I came up with an elaborate plan that would have knocked my own socks off.

My idea was a scavenger hunt that would take Alana to five different locations, leading up to a big reveal at the end of the night. First came the ruse: I had my client tell Alana that his birthday gift to her was a night out with her best friend. That night, a big black limo arrived to pick up Alana, and inside was her best friend, who was holding the first clue in a sealed envelope. The answer to that clue took Alana and her friend to the first location, where they met two more of her friends, holding a second clue. At each new location they were ushered to a special table or a private room and given meticulously chosen cocktails and appetizers, and Alana continued to gather more girlfriends—all of whom were in on the surprise.

Finally, the last clue led Alana and all her friends to the Rainbow Room, where I met her, pretending to be the hostess, and led her to a private room. The planning of the whole evening had been so elaborate, and my client's excitement and anxiety level had been so high, that I felt my own heart beating faster as we approached the pocket doors that led into a private room. As I pushed them open, Alana was met with a shout of *Surprise!* from everyone she loved most in the world—her boyfriend, her family, and a slew of out-of-town guests. As I stood back to let the cheers wash over Alana, I felt the residual glow myself.

My client was beaming. Here was this highly successful man who had people catering to him every day of his life, but underneath it all he just wanted the woman he loved to be happy. It was a bittersweet moment for me. This time I hadn't planned the perfect night for a client—I'd planned the perfect night for me. But it wasn't me in that doorway, meeting the embrace of everyone I loved most in the world—it was someone else. It was *always* someone else. I felt like a young kid at a friend's party, watching her opening up birthday presents, unable to hold back my jealousy that she got the gift I had always wanted. At one time, I'd believed that standing on the sidelines of other people's events was all I could expect from life. That by osmosis these events would bring me the emotional satisfaction they did for my clients. And for a long time, they did. But somewhere inside, I was starting to hope for more.

Around this time I was able to buy out the original investor in my company, and now Save the Date® was one hundred percent mine. It felt like a divorce, and I had been awarded sole custody—and along with it came all the financial and emotional responsibilities of being a single parent. It was a huge step for me, and the process was wrenching and stressful, but it also felt like a fresh start.

On a whim, I decided to apply for the Entrepreneur of the Year® Award sponsored by Ernst & Young. It was a lengthy process, including essays, followed by a series of interviews. Not for a second did I think I had a chance to win, but it was a vote of confidence in myself to try. Why not? I didn't care that I would lose.

When I was named a finalist in the under-forty category, I laughed out loud. The other finalists were men and a good ten years older than I was, and their companies were technology start-ups with hundreds of employees. Meanwhile, I was twenty-nine, I had six employees who were women my age or younger, and I was a *party planner*. Give me a break—there was no way I was winning that award. I figured they'd only included me for diversity reasons.

The awards dinner was held in a massive banquet hall at the Marriott Marquis. There were at least 1,500 people there, and while the other finalists had multiple tables, I just barely managed to fill one table with my staff and my parents. There was a ridiculously massive screen in front of the banquet hall so they could air documentaries about each of the finalists. In mine, I seem to recall that they interviewed a number of my clients, but the worst part for me was when I was the talking head up on that movie-theater-size screen. The documentary described how I had shifted the paradigm of my business and turned the whole structure on its head. I should have been feeling on top of the world, but instead I felt like such a fraud, competing with a crowd of MBAs and captains of industry with my measly little event-planning business.

When they announced the winner—Jennifer Gilbert of Save the Date®—I was so shocked that I just sat there, frozen. Finally Nikki, the staff member who'd been with me the longest, said, "Jen, get up!"

My spotlit walk up to the podium was an out-of-body experience. Standing at the podium was surreal, and looking out at those 1,500 faces was so unexpected that I hadn't even prepared a written thank-you speech. The only thing that settled my nerves was to look at the table where I'd left my staff and my parents. I saw six young women full of shock and joy for me, and I was struck for the first time with a startling fact: I had never thanked them. All those years of wearing my scary mask, I hadn't been able to show them the gratitude I felt in my heart every day. Finally, I found my words. I think I started by saying that it was the first time in my life I'd ever been speechless. And that most of all I needed to thank my employees—the people who showed up every day *for* me and *with* me, and who had made my company and that award possible. It was as if someone had turned on the lights. These women were my constant well of reinforcement. That office and my job fed me all those years when I felt small and unimportant and needed external validation.

That night marked a major turning point for me, and the beginning of a process that would change the entire way I looked at my company as a business owner. There were multiple steps along the way, and the next came soon after the awards dinner. Nikki, who'd been the one to nudge me up to the stage, was a tall, gorgeous South African woman, and no shrinking violet. She was my first real employee, and she was the one all the newer employees looked to for advice and guidance. When she came knocking on my door one afternoon and asked if she could speak to me privately, my heart did a little jump. I felt nervous—*Oh my God, was she leaving?* I could sense that she had something serious to say. But my way of dealing with nervousness was to shut down emotionally. My face got stern, and I said, "What's wrong?"

Nikki sort of laughed, but I could tell she was taken aback. She said, "You know, Jen, after all these years you still scare me."

Me? Scary? I knew I could be intimidating to the younger women in the office, but I'd known Nikki for years at this point. Hadn't she been able to see past my scary mask? And if not, then I really *did* have a problem. I quickly apologized to her for freaking her out, and then she sat down and told me a story. She said she wanted to thank me, because she'd just experienced a stressful situation in her personal life, and she'd gotten through it by hearing my voice in her head. She said that when she was nearly convinced that the situation was doomed, she heard me saying, "You don't ask, you don't get, so go get it." And then she found herself a yes—and she wanted to thank me for that.

For the first time in all my years of working, I burst into tears in the office. I was moved that I had been a role model for someone else. Me—the inner basket case who for so long had felt like she was barely holding it together. This was an epiphany for me. It gave me a glimpse of the whole, integrated person that I wanted to be: a person who carried her beliefs, her kindness, her love, and her passion into every area of her life.

After the trial, I'd realized that I couldn't go it alone in my personal life anymore. Now I'd reached that point in my business life as well. I made a vow that I would never walk into my office again if I couldn't smile and offer a salutation of some sort. If it meant that I needed to get a cup of coffee at the corner deli and walk around the block a few times until I could shake my mood, then that's what I would do. I didn't want to be scary Jen anymore. I didn't want my people to be frightened of me, waiting to see what kind of day they were going to have based on my mood that morning, and I never again wanted to sit in my office feeling like a freak because I found it so challenging to say a simple hello. I cared about my employees as people—their home lives, their happiness in and out of the office, and their personal growth—and it was time to start showing them.

But I knew there was more involved with transforming my company than making resolutions. I realized that I'd taken the company as far as I could on my own, and now I needed help.

I had heard about a course at MIT called "Birthing of Giants." A master's program for entrepreneurs, it was an intensive series of classes in all the high-level stuff that I'd never thought about in a concrete way—vision statements, corporate culture, best practices. This was exactly what I needed, so I applied and was admitted along with sixty-four other under-forty business owners from around the world (sixty-one were men; one of the two other women was part of a husband/wife team). I worked in a female-dominated industry, my employees were women, and I've always thrived on my relationships with women. Not only was this totally male-dominated environment alien to me, but I felt deeply intimidated by the other students' knowledge. Most were MBAs with all kinds of business expertise that I'd learned by my wits—not in graduate school. I felt like such a fraud. Here I was, just having won an award most of them had applied for and lost, and I had already forgotten about that accomplishment.

So I put on my tough exterior armor and sat at the front of the class, absorbing the course work like a sponge but speaking to no one—at least initially. When the others were getting together for drinks in the evening, I was back in my dorm room, poring over what I'd learned during the day. I'm sure I was known as "that bitch from New York."

Over time, though, I started to warm up, and I was deeply affected by the passion of my fellow students. There was always an incredible array of speakers, and each of the business owners who attended was invited to tell the story of his or her business at some point during the course. I listened while these strangers spilled their hearts about what their businesses meant to them, and how they had poured so much meaning into their work. For the first time, I felt surrounded by kindred spirits. I realized a truth that has stuck with me ever since: *Everyone's got their something.* Everyone in that room had a story—whether it was sickness, poverty, divorce, or some other adversity—and they had all channeled their personal challenges into something beautiful. Their stories might be different from mine, but we all had one.

Finally, on the last day of the course, I was the only person who hadn't spoken. This was at a time when the people who knew what had happened to me were a very select few, and certainly no one in my office knew. I'd never sat down and told a bunch of girlfriends what had happened, much less sixty-four strangers whom I'd been so intimidated by just a short time before. In telling me their stories, these strangers had shown me the respect of treating me as their equal, as if I was as worthy as they were to sit in that room.

I got up in front of them, and for the first time in my life I told a large group of people about my personal tragedy, and how my company had been born of my commitment to spend the rest of my life helping people celebrate. I said that I got up every day and helped people to laugh and express themselves, and I loved what I did. Their response was staggering to me—a standing ovation followed by dozens of e-mails telling me how much my story had meant to them.

It was a life-changing experience for me to reveal myself that way among peers, and to feel nothing but respect, acceptance, and gratitude in response. It taught me that at least some of the time I could fully be myself—all the sides of me present and visible for the world to see.

I took everything I learned at MIT, and I brought it back to my company. The first thing I did was to change my title to chief visionary officer; aren't events, and business, and life, all about vision? Inspired by my experience of telling my story, I decided that it was time to come clean with my people about who I was and why I believed so strongly in the work that we did. I gathered all my employees together for an off-site retreat, and we talked about our mission statement as a company and why they thought our work was important. And then I told them my story, every bit of it and then I said, "I'm sorry for not expressing my gratitude for all those years." I broke down and cried right there in front of them, looked around into caring, tearing eyes. This time, I wasn't making them uncomfortable or nervous; they felt compassion for me. It's amazing how disarming a simple apology can be, not an "I'm sorry but" which feels conditional, but an honest "I'm sorry," which is limitless. I realized it was no longer enough to work hard and expect everyone else around me to do the same. We all needed to know why we were there, and why our work mattered. Of course I also hoped that they might gain some understanding of—and maybe a little forgiveness for—the scary mask I'd worn for so long.

Not long after, my staff was planning a massive four-day event for an international nonprofit organization. This was one of the most important clients we'd ever had, and the conference involved hundreds of leaders flying in from all over the world, multiple venues, meals, conference rooms, and daily activities for the leaders as well as entertainment for their spouses. The finale was a dinner cruise around Manhattan. We worked like dogs to make sure that everything was perfect, and the whole four days leading up to the cruise had been seamless. I trusted my people implicitly, and I gave them positive feedback for their work. It was proof that I could be both a perfectionist and a human being.

Then on the last night, as the dinner cruise approached the dock to unload my clients after their perfect evening, a pin in the brake pad broke. The boat bore down on the dock, unable to slow down or stop until it actually crashed into the mooring. I mean, this was not like a little *Oopsie, we just bumped the dock*. This was a plate-smashing, people-flying-in-the-air collision with the dock. I couldn't have invented a worse ending. Luckily the injuries were only minor, but the finale of my four-day extravaganza was sirens and ambulances instead of smiles and goodie bags. It was an unmitigated, unpreventable disaster.

When my staffer called me to tell me the news, I could sense how terrified she was. I don't know what she expected or feared I might say, but she knew how much was riding on this event. She knew not only that this organization symbolized millions of dollars of business but also that our reputation was on the line. My reaction to her in that moment would be the test of the new me. Who would I be in a crisis—the scary boss who prided herself on her supernatural perfectionism, or the woman with a mission much larger than any single event?

I passed the test. I reassured her that there was no way any of us could have predicted a broken brake pin (I mean, *seriously*? Did I even know that party boats had brake pins?). And I told her that the most important thing was that she was okay. And I meant it. From the bottom of my heart and soul, I meant it.

CHAPTER ELEVEN

I Never Promised You a Goodie Bag

We all want a happy ending. If every event is a story, then a goodie bag is like the kiss at the end of the fairy tale. We're all waiting for it, and when we finally get our prettily wrapped reward, we tear into it like a child at a birthday party.

Inside that goodie bag there might be nothing more than a granola bar and a shampoo sample, but party guests will sharp-elbow their ninety-year-old grandmother out of the way if it looks like there might be a shortage. You think you've seen the worst humanity has to offer at a Prada sample sale? Try imagining the last lifeboat on the *Titanic*. The last helicopter out of Saigon. Now imagine me standing between a horde of upper-crust revelers and the single remaining goodie bag. I swear to you: it will get ugly. Grown women will fight over who got the better shade of lip gloss. Guests will scream at my staff because they want to know why they got the regular goodie bag instead of the VIP goodie bag. They will actually climb over each other to get to those little bags, and then they'll end up tossing all of it. So many times I have walked out of events to see trash cans full of the books, pens, and picture frames that my staff and I had spent hours stuffing into those shiny bags.

At most events, the bags aren't put out until the very end of the night, and I have seen guests actually leave a party early just so they can stand on line to wait for their goodie bag. Every time it happens, I'm surprised all over again. *Really?* I want to say to them. *You left the good time in there, just for the promise that there might be something worth having in this little bag?*

Once I did an extravagant party for a massively successful biotech company. The invite went out in a black box, and packed inside were hangover cures, party poppers, confetti, all to foreshadow that this would be the event of the year. At the party there were aerial performers and seven-foot-tall ice sculptures that shot vodka into custom-made glasses in the design of the company's newest prescription launch. For the unveiling of the new drug, the CEO was lowered down among his one thousand guests from a high, darkened ceiling. The whole event was like a circus, a game show, and Mardi Gras rolled into one. Finally, after this incredible star-studded, million-dollar event, a guest came up to me and said, "Where are the goodie bags?" I had to laugh.

Over the last few years I've conducted my own secret war against goodie bags. I've gently pointed my clients away from the pens and the lip gloss. If they really feel they must give something, then I'll encourage them at least to make it fun and edible—cookies and milk, or the next day's *New York Times* and a bagel. When I planned a big banquet for a women's nonprofit, I managed to convince them to make a donation in lieu of dessert and goodie bags. At all events, everyone takes dessert being served as their signal to leave. So instead of wasting all that food and spending money on goodie bags, my client announced to the guests that they'd donated the dessert course to a shelter for women. I loved that gesture, because it acknowledged the meaning behind the event. After all, life's not about the presents, but your presence. I wanted to jump for joy and scream, "Rock on with your big bad selves for paying it forward!"

To me, goodie bags have become a metaphor for life. It's fine to hope for a happy ending to any given situation, but it's when we expect a really specific outcome that we can get into trouble. Yet, truth be told, I had gotten to a point in my own life where I was kind of looking around and saying, "Is this it? Where's *my* goodie bag?" In my case, that goodie bag was "the one"—my soul mate.

During the years that Rachel, Deanna, and I were living together, we were all single at one time or another— often at the same time—and we came up with a brilliant idea that I swear merits a patent. We'd throw a huge party and invite all of our friends, colleagues, and basically every-one we knew, and ask them to bring their cutest unattached male friends. It was genius—a party where at least half the people would be single men. How often does that happen?

Our building had a roof deck that everyone could use, and we packed it full. My specialty was something I called sangria but that was actually clean-out-your-liquor-cabinet punch mixed up with some fruit and sugar. I served it in enormous bowls, and by the end of the night everyone was cross-eyed.

Those parties were a buffet of men, and without fail we'd all meet someone before the end of the night. When the party was over, there would be seven of us left: Deanna and Rachel with their potentials, me with my new interest . . . and of course Bennett, washing the glasses. He was always there, the pal who'd take the trash bag when I handed it to him as I went off to flirt with someone else. And he never once com-plained. One year, when I was already paired up with a boyfriend, I remember the man saying to me, "What's with Bennett?" I just laughed and said, "He's my sidekick and my best friend."

On one particularly memorable night, I was juggling two men that two different friends wanted to set me up with. Before the party, my friend Susan had been talking up a man named Evan. She said she had a feeling we'd really like each other. He was a Harvard Business grad who was running his family business and had an entrepreneurial streak. And he loved music and dancing, which were my obsessions. So I said, Fine, fine, bring him. Then my friend Lisa wanted me to meet a single friend of hers. He was gorgeous and incredibly charming, and forever after my friends and I would nickname him Beautiful Boy. I was never afraid to stack the decks, so I told Lisa that of course she should bring him. The more the merrier.

At the party, from the second I laid eyes on Beautiful Boy, I was mesmerized. I happily neglected everyone else. Toward the end of the night, I was approached by a cute boy in glasses and a funky plaid shirt. He said, "Susan will kill me if I don't at least say hello and thank you for inviting me." I looked at him blankly, and then he said, "I'm Evan, Susan's friend." Then I remembered: Evan, of course. So I said, "Oh! It's you, hello, thanks for coming." Evan invited me to go out dancing with his friends, but I begged off and said I had to stay with the party. Then I happily went back to talking to Beautiful Boy, whose actual name was Dave. He was the last man there (not counting Bennett), and before he left, he asked me out for Friday night. So of course I said yes.

The next day Susan called me and said, "So what did you think of Evan?" My eyes were filled with Dave, and I could barely remember Evan. I told Susan I thought he seemed nice. Then she laid it on thick—*Jen, I swear he's great, he thought you were so pretty, he really wants to see you.* (Meanwhile she pulled the same routine with him, I'd find out later—*Jen thought you were adorable, etc., etc.* She played us both like fiddles.) I said, Sure, have him call me. And he did—so we made plans for that weekend. But really, my priority was my date with Dave.

My first dates with both men went exactly the way I might have predicted. Evan picked me up in his bright yellow Mazda Miata (which I loved and we bonded over—not just anyone who's not a cabbie can rock a yellow car) and took me to a cool Moroccan restaurant. He was too nervous to eat, but I got him to loosen up a little bit by asking, "So how was your day?" He kind of looked at me, a bit surprised, and said, "What do you mean?" I said, "I mean, how was your day?" To me, that's a more caring way to find out about a person's life rather than to just up and ask them what they do for a living. So Evan told me about a problem he'd had at work that day. Then he said, "You're a business owner, what would you have done?" I said, "Hmm. I would have handled it completely differently." And then I gave him my opinion. Evan started laughing and said, "Uh, why don't you tell me how you really feel?" I said something that's my motto to this day: "If you want a different answer, ask a different girl."

During our date, I noticed that Evan took me seriously, and I liked that. On the way home we talked about music, and I realized that he knew all the same obscure hip-hop bands that I did. It was a fun evening, but there were no sparks. We shook hands at the end of the night, which was totally typical for me.

Dave was a different story. We flirted all evening during our first date, and I cannot for the life of me remember what we talked about. And I think we made out in the cab. Well . . . rules were made to be broken.

The next weekend they both asked me out again, and I said yes again, although I was still putting my money on Dave.

That Friday, Evan picked me up at my building, and he hailed a cab. I remember I was wearing a halter dress and these funny green butterflies in my hair (not sure what that fashion moment was about, but anyway), and when we got in I leaned forward to tell the cabdriver what route to take to the restaurant. It was laughable—even on a date I couldn't loosen my type A control-freak personality. Meanwhile, Evan must have been thinking, Who is this pushy broad?

While I was talking to the driver, Evan's only view of me was from behind. He was quiet while I gave my instructions, and then he said two words to me that shifted the whole dynamic. Those two words were: "Nice back."

I don't know exactly what happened in that moment, but it was certainly chemical. There was no other way to explain it. I turned around to look at Evan, and that was it. I thought, Oh my God, I love him.

We went to Grove Street, a romantic bistro in the West Village, and sat in the outdoor garden. I looked across the table at him, and I was thinking, How did this happen, could I already be in love with him? All my senses were firing. I knew that something was going on here—I had recognized something familiar in Evan.

I said, "What's the story with you, what's the issue? There's something a little broken in you, I can feel it." I guess it takes one shattered spirit to know another, and in the middle of dinner he opened up to me about his complicated family relationships, old wounds that were magnified by the fact that he worked with his father and uncle in the family business. Meanwhile, he really wanted to leave and make his own success, but he felt a tremendous obligation to carry on what his grandfather had started, and he was pretty resentful about all of it.

I listened to him, and then when I got up to go to the ladies' room, I did something so uncharacteristic of me that I'm surprised at myself even in retrospect. I walked around to his side of the table, I took his face in my hands, and I kissed him—deeply—right in the middle of that tiny thirty-person restaurant. It went on so absurdly long that someone actually picked up a piece of gravel from the ground and threw it at my back—like, *Enough, get a room.* I was so embarrassed that I abruptly stopped kissing Evan and ran into the bathroom.

I'm sure he was as shocked as I was, and when I got to the bathroom, I was shaking all over. I took a few deep breaths before I opened the door to leave, and there he was, standing right outside. I said, "Do you have to pee or do you want more?" He came in with me, and we stayed in that three-by-three-foot bathroom for a good fifteen minutes. Just kissing— but still. I didn't do things like that.

When we left the restaurant, I sat down on the stoop of a brownstone, and I said, "I have to tell you something." And then I told him everything—the whole story of the attack—which I had never, ever done on a second date. He dropped down next to me, and we talked until 4:00 a.m. He took me back to my apartment, and then he went home.

At nine the next morning my doorbell rang. I opened the door, and there was Evan with two dozen roses.

I never saw Beautiful Boy again.

After Evan and I had been together only two weeks, I was going to Italy for a wedding. I'd be gone for ten days, and I managed to convince him to join me for the last weekend of my trip. Before I left, Evan wrote me a sonnet that was hand-delivered with a bunch of sunflowers. He drove me to the airport (something that to this day always makes me feel special) and packed me up with seven presents—one for each of the days we'd be apart. Meanwhile, I'd ordered personalized cookies to arrive at his office after I left. We were each competing for who could outromance the other, and it was intoxicating and explosive. Very quickly we spent almost all of our time together. A few nights a week I'd peel myself away to spend the night at my own place, but most often we were at Evan's.

Once I planned a lavish and complicated wedding for the daughter of a fashion mogul in the Hamptons. It was a ceremony on the beach with torches set up everywhere and a wedding arch that was handmade from tree branches intertwined with wildflowers. The bride didn't want any of the guests to see the reception room prior to the wedding, so we had to create a cloth-draped, arched hallway that led guests through the dining room of the club and out to the beach. Not only did we have to construct the hallway, but we had to shove all the furniture aside to accommodate the hall, and we had to curtain all the windows as well so that no one could see inside. Meanwhile, there weren't enough chairs for the ceremony and the reception, so after the ceremony was over, we had to quickly grab all those chairs and bring them into the dining room. Usually we'd have an entire day to set up a room, but in this case we had exactly an hour and fifteen minutes between the ceremony and the reception. During that time the guests would be escorted to a long deck where cocktails would be served and the father of the bride planned to saber a bottle of champagne (literally—he used a sword to open it).

While the two hundred and fifty guests were drinking, we not only had to set up the (two hundred and fifty) chairs and tables, but we also had to create the custom-made dance floor that had been painted with the couple's new monogram. The dance floor had been cut into moveable pieces, and each piece was numbered so that it could be assembled the right way—like a huge forty-by-forty-foot jigsaw puzzle.

I had an army of people with me that day, and we all worked like maniacs—the drama that the bride wanted to create for her guests meant all kinds of drama for us. The whole day was exhausting and physically intense—it started at 7:00 a.m. and didn't end until 2:45 the next morning, when we finally finished breaking everything down. By the end of it I had splinters in every finger, twigs in my hair, and my legs were cramping from standing in high heels for nearly twenty-four hours straight. The bride offered to put me up in a room that night, but all I wanted was to get home to Evan. So I called him before I left and told him not to wait up, I'd crawl into bed when I got to his place—I just wanted to wake up next to him in the morning. Driving home, I was practically blind with fatigue. I expected Evan to be asleep in bed when I got there, but instead, when I walked into his apartment I found him still up and waiting for me, with candles lit all over the wildflower-strewn living room. It took my breath away.

At first it felt like Evan and I had everything in common. He loved my success and that I owned my own company, and he was the first boyfriend I'd ever had with whom I could really talk business. At the same time, he loved my silly, goofy sides that had all but disappeared, and I was able to relax again. We made up our own rap songs and laughed like kids. There was something magical for me in that—it made me feel innocent again. And when he made himself vulnerable to me, and shared his pain and insecurities, I felt a powerful urge to help.

He was romantic and brilliant, and he made me feel smart *and* beautiful—no man had made me feel both before. While I'd never been the dreamy type, suddenly a part of me started fantasizing about the future again. I pictured a big wedding for us, the kind of perfect fairytale event that I planned for other people. I remember driving with Evan in his car one day—the sun was warm and shining, and my heart felt light. I thought to myself, I'm in love, and life's good, it's just that simple. It was a feeling of such optimism.

I was coming out of my emotional fog at just the moment I met Evan, and there he was, standing in a ray of light. The timing was finally right for me. I didn't want to hold myself at a distance anymore—I was primed to fall in love. But Evan was in a very different place in his life, that same undeserving place I knew all too well.

He thought after he graduated from Harvard Business School he would take over the world. Instead he'd taken over his family's broken business. He was trapped by his unrealized expectations, and no amount of unconditional love from me could rescue him. I told him that he was brilliant, and wonderful, and I believed in him. But he didn't want to hear that. He told me that he expected me to be his mirror, and that I should be telling him the unvarnished truth. It was as if he wanted me to kick him when he was down, because that's what he thought he deserved. I had learned over the years that you can't heal—or forgive—until you give up the struggle. But Evan couldn't give it up—he was still too inside of it.

About two years into our relationship, Evan's frustration at work had blown up into desperation, and he seemed to be manically flailing around, searching for an escape hatch. Once he asked me to go along with him to a convention he had to attend in New Orleans. While we were there, he seriously raised the idea that he would move there and live on a houseboat. I said, "Are you kidding me? After two years together, you want to give it all up and move onto a boat?" His response was, "You can visit."

He gave up on the idea, but my confidence was rocked. I started to feel that desperate ache to prove something, and this time I was going to fight to hold on to love, at the expense of my heart.

It didn't occur to me that I shouldn't have to work so hard to make him happy, or that love wasn't a fight. Yet still I couldn't let go. For the first time in my adult life I was wholly and completely in love. All my years of learning from my mistakes and being willing to take new turns had taught me the very big lesson that love *was* the answer —not money, or appearances, or pride. Now I was finally ready to embrace love with a grateful, open heart. Unfortunately I had chosen the wrong person to throw my arms around.

. . . .

Soon after New Orleans, Evan broke up with me. No matter how much I was trying to *will* this relationship to work, I couldn't control another person's actions. I cried so long and so hard that my face swelled up and I couldn't open my eyes. Every wedding I planned felt like tearing off a bandage. None of my old coping mechanisms were working, and I was a disaster—just a complete and utter disaster. I begged Evan not to call me unless he was ready to make a real commitment, because otherwise simply hearing his voice would destroy me.

Eventually I had to pull myself together—not just for my own sake, but for everyone else's. To stay sane, I knew that I needed to keep myself really, really busy. To get me out of my house, Bennett started planning an outing a week to try a new restaurant, each in a different neighborhood. Then I received a call; a friend of Nicolette's from Norway was coming to the city, and I promised to show her around. Marianne was the perfect distraction for me, and every night we were out. She was tall and gorgeous and game for anything, and I dragged her around to every hot club in the city. It was hard work, though. By the end of at least several nights, my party-girl facade couldn't hold up any longer. Emotionally spent and not a little drunk, I'd show up at Bennett's door, ringing his bell at 3:00 a.m., screaming into his intercom, "*I need chocolate!*"

Bennett would answer the door and lay out a buffet of sweets—he was one of the few who knew that despite all outward appearances, sugar was my drug of choice. I'd gorge myself to the edge of nausea, and afterward I'd pass out. The next morning I'd wake up on Bennett's sofa—hung over, with a dribble of chocolate and caramel down my chin. He'd just shake his head as he strolled through the living room on the way to get his morning coffee.

I received two letters from Evan during that time. He told me he missed me every day, but he never said the exact words that I needed to hear—that I was "it" for him. He did say that he was trying to be the man I wanted. But even I knew that trying and doing were two different things. It killed me, but I didn't respond.

To make matters more complicated, while I was in deep mourning for my relationship with Evan, my sister Rachel was planning her wedding. Rachel had gotten me through my absolutely darkest days. Now she was in love and getting married to Daniel, and I was so happy for her. I was incredibly honored to be her sister, her maid of honor, and her wedding planner rolled into one. I put all of my now highly potent wedding juices into planning her big day.

Yet I couldn't help despairing a little when I compared her life with mine; I was the older sister by three years, and it was all happening to her first. On one particularly difficult morning after my breakup with Evan, I was scheduled to go dress shopping with Rachel, and it was exquisite torture. I stood in the Vera Wang bridal boutique, watching her try on one dreamy dress after another, and I could not have been more miserable, and more convinced that there was no happiness in this world for me. I even remember what I was wearing: overalls (God help me) and a wife-beater T-shirt.

Right after Vera Wang I went out to brunch with Bennett. I must have been a real joy to behold that day in my overalls, not a stitch of makeup on my face, my hair in a ragged ponytail. I think I was actually wearing a scrunchie (remember those?). I was in such a lousy mood and feeling so alone that I was generally sick of myself. I said to Bennett, "Let's hear about you. Who are you dating?"

He said, "No one special."

I said, "Honestly, Bennett, you are the pickiest person I have ever met. I wouldn't even know who to set you up with."

Then he said, "How about you?"

Whoah. What? I looked at Bennett, and I blinked—once, twice. This was not *at all* the direction I expected this conversation to take. I was speechless.

Bennett proceeded to tell me that he'd loved me for six years, and stood by and watched me through one boyfriend after another. He told me that he couldn't wait another day to declare his love because he knew from experience that I'd have another boyfriend by then. This was his chance, and he wasn't going to pass it up. He said that he would love me forever, and we were soul mates, and that he was certain that he wanted to marry me.

It's a cliché to say that I fell out of my chair, except in this case I really did fall out of my chair.

After I picked myself up from the linoleum, Bennett just sat there quietly, waiting for my response. I looked back across the table at this man I adored, and I thought: He's my rock, my confidant, my best friend. I had shared every painful life experience I'd ever had with him. The only answer I could give him was: "I'm unbelievably flattered, but I'm in love with someone else. I don't feel that way about you, and it's just never going to happen." And I also think I said, "And I can't picture kissing you."

He'd spend the next two years trying to change my mind.

PART IV

The Best Man

Unless you walk out into the unknown, the odds of making a profound difference in your life are pretty low.

—TOM PETERS

CHAPTER TWELVE

Best Friends

It's not easy planning weddings when you're unhappy in love. You can't help wondering what you're doing wrong when all these people around you are arriving at a destination that seems wholly out of reach for you. But, oddly enough, it was a wedding that helped me say my final good-bye to Evan.

While Evan and I were still dating, I'd met Jenny, his best friend from Harvard Business School, and her boyfriend Mark. Everyone knew the two of them were destined to be together, but Mark was having a hard time committing. But after some ups and downs, they decided to move in together prior to getting engaged. Jenny was ecstatic, and Mark was genuinely happy, too. To me, Jenny seemed to have it all, and I couldn't help feeling a little jealous.

Jenny's bliss was tragically short-lived. Just when she should have been focused on her new life with Mark, she learned that her mother had been diagnosed with stage-four cancer and given only a few months to live. Jenny called Mark, crying hysterically, and said, "Mark, I'm so sorry to ask this of you, but we have to get married now, my mother has to see me get married." Although Mark had been the one to take to their engagement slowly, without a moment's hesitation he said, "I'll take it from here, Jenny." He proposed ten days later, after finding the perfect ring and writing her the most heartfelt poem. I was awed and moved by the entire story.

I was heartbroken for Jenny, and I wanted to do something. I'd learned over the years that when someone is suffering, you don't ask what you can do—you just do. So even though I barely knew Jenny, I called her, introduced myself as "Evan's girlfriend Jen," and told her that I couldn't do anything to ease the pain of her mother's news, but I was a party planner, and I could plan her wedding in three months so that she could have her mother by her side. It was my gift to them, and it meant a great deal to me that my work could bring her and her mother solace, and maybe even a little joy, at a time that seemed bereft of both.

It was soon thereafter that Evan and I broke up. Since Jenny was Evan's good friend and not mine, our breakup might have made it awkward for me to continue with the planning. But over the course of working on her wedding, I became extremely attached to Jenny and her family. It was a deeply emotional time, and it brought home to me how important our rituals are—a wedding isn't just a big party, or an opportunity to buy a new dress. It's a momentous occasion, and the bond that Jenny and her mother shared during those weeks made it feel all the more significant. Jenny's wedding would be a beginning and an end—a blessing and a good-bye. Each week Jenny's mother got thinner, and the time they had left together felt shorter, and they treasured every second together.

Over the course of the weeks while I was planning Jenny's wedding, I was in my frantic phase of trying to put Evan out of my mind—while secretly hoping that he'd come to his senses, just as Mark had. But aside from Evan's two noncommittal letters, nothing had changed, and I hadn't heard anything else from him. Then I heard from Jenny that Evan had requested that I be seated next to him at the reception. I was horrified—somehow I'd managed to forget that of course I would have to see Evan at the wedding. It was some kind of cosmic joke—me and Evan, *at a wedding*?

I told Jenny that, as much as I loved her and was so honored to be her planner as well as an invited guest, I just couldn't bear to see Evan, and therefore I couldn't possibly come to the wedding. Then Evan himself called me—for the first time since our breakup—and pleaded with me to reconsider. Finally the guilt of abandoning Jenny on her big day won out over my sense of self-preservation, and I agreed to go to the wedding, on the condition that I would not sit with Evan. I also agreed to meet Evan for a drink. I told myself that it would help to defuse the drama of seeing him at the wedding if at least it wasn't the first time since our breakup.

It was three days before the wedding when we saw each other. He told me he loved me and missed me. I was guarded, but not guarded enough.

Jenny's wedding was at a loft in Tribeca, and it was elegant, simple, and purposeful. The flowers were all white, and there was a fantastic soul dance band, because the bride and groom both loved music. Jenny and Mark had put a great deal of thought into their vows, and every aspect of the ceremony felt infused with emotion. Of course every wedding should feel that way, but so much of the time we get more focused on the color of the napkins than on the words that we say to the person we're marrying.

During the ceremony, tears rolled down my face and formed dark drops on my dress as I thought about what Mark had done for Jenny. He'd overcome every petty hesitation he had, and he'd thrown himself into making this day everything that Jenny wanted it to be. The most important thing that Jenny wanted was to have her mother there to see the day—and she was, wearing a pretty wig and a lovely dress that we had to keep taking in around her thinning frame in the weeks before the event.

As I watched the wedding unfold, my eyes still swimming in tears, I knew that Mark was a man in love. Marrying Jenny in three months wasn't a sacrifice for him. Seeing his fiancée in pain and knowing that he could do something to help—he just did it, and without a second thought.

I looked over at Evan, who was in the wedding party, and his face was tear-streaked, too. After the ceremony he told me that he loved me, and that I was it for him. It wasn't a marriage proposal, but I heard what I wanted to, and I think in that moment he genuinely believed what he was saying.

After the reception I said good night to Evan, and he looked shocked that I was leaving without him. We'd spent the whole night dancing and had a wonderful time together, and he thought that was it—we'd go back to the way things were. I told him that I couldn't act as if nothing had happened, and that if he wanted to see me, he should call me, and we would talk.

The next day he called, and we spent the entire day together on his boat. I cried and cried. I told him that he'd devastated me, and that I was terrified he'd do it again. I told him that I didn't need a ring from him—it had never been about that. I just wanted to know that I was the girl for him. I asked him if we'd be together forever. He cried with me, and he said, "That's the goal."

In the weeks after Bennett had confessed his feelings, I worried it would be awkward for us to be together. But of course, since it was Bennett, it wasn't. We still spent time together, avoiding the topic of *both* our love interests, and just enjoyed each other's company. But now, even though I wanted to protect Bennett and try to spare him the pain that Evan and I were reunited, I knew I had to tell him right away.

I told Bennett that I was sorry, and that I loved our friendship, which over the years had become one of my most important relationships. But I also told him that if he needed to break up with me as a friend for a while, I'd completely understand. Never for a minute did I want him to think that I was stringing him along in any way. As his friend, I wanted to tell him, *Screw it, that girl doesn't appreciate you, walk away.* But since I was the girl in question, I didn't know how to handle it. So I said the truth: "Bennett, I love you way too much to risk losing you for the sake of 'romantic love.'"

I didn't hear from Bennett for a while after that; then he e-mailed me one day and said, "You're still my best friend and I want you to be happy. If you love Evan and you're going to marry him, then I'd better get to know him a little better."

Bennett knew that Evan loved hip-hop, so he bought the three of us tickets to see *The Bomb-itty of Errors*, which was a hilarious mash-up of Shakespeare's plays set to hip-hop music and dance. It was clever, and perfect, and so thoughtful. And with this one gesture it showed me he was my truest friend. Despite his own desire and hopes for himself, he wasn't making me choose; he would let me have both kinds of love.

One day I was walking by a jewelry store, and I noticed a really cool ring in the window. It was Gothic and funky, and I mentioned in passing to Evan's sister how much I liked it. Then I put it out of my mind.

Our third anniversary was approaching, and we'd planned a trip to the Greek Islands to celebrate. Before we left, Evan presented me with a small, ring-size box, and my heart did a little leap. Even while my pulse was fluttering, Evan laughed and said, "It's not *the* ring, but it's the promise." I opened it, and it was the funky ring that I'd admired in the window. I loved the ring, but my heart sank a little, and I tasted the all-too-familiar tang of disappointment.

In Greece, while sailing around the islands, I made what I thought was a major concession. I suggested to Evan that we think about moving in together. I'd always said that I wouldn't live with a man before we got married, but my whole relationship with Evan had been a process of slowly breaking all my own rules, and I was willing to make the compromise for him, if that's what he needed.

We continued talking about renting together after we got back to New York, but in the meantime I'd decided that I really wanted to own something. I got fixated on buying a weekend house in Rhinebeck. I was thirty-one years old, and I felt like I'd been working a zillion years, and I wanted a house to show for it.

I wanted to believe that there was still a future for Evan and me, so I asked him to come with me to look at houses. I thought, When we end up together, it will be nice to have his input. We were touring around with a broker, and there was one house in particular that was quirky and completely impractical, but we both adored it immediately. Then Evan said, "Let's do it, let's buy this house together." We got so excited we even named it. It was the "mushroom house." It did feel a little like putting the cart before the horse—but I told myself that buying a house could be viewed as a kind of commitment. I consciously overrode my own instincts.

With Evan, I'd allowed myself to dream of a fairy-tale happy ending. The thought of letting that go was devastating to me. It would be so much more than breaking up, it would be *giving up*. Eventually, though, I couldn't ignore my instincts any longer. Although I was terrified of what it might mean for my future, I had to admit to myself that Evan was no more ready to commit to me now than he'd ever been. The plans and possibilities of that house were moving us forward, not his certainty about me. At Jenny's wedding, I'd seen what real love and commitment were, and what Evan and I had was only a shadow of that. After yet another Saturday afternoon of going back to look at the mushroom house, I asked him to pull over the car. Tears were streaming down my face, and he looked at me questioningly. I said, "What are we doing?"

Four simple words, but they cut right to the heart of the matter. We weren't doing anything, and we weren't headed anywhere. It was finally over—for good this time.

I went into a hole for a long while after that. I was so tired of fighting for everything. I thought that there must be people out there who found love and a life worth living without needing to work so hard all the time. I wanted that easier way, but I still couldn't surrender the fight.

I had stopped seeing my therapist, Ann, but I knew that I needed her help again. I was back on one of my circular tracks, and I was afraid that without a little outside guidance I was going to keep on making the same mistake again, seeing a future where there was none. What's the definition of an idiot? Someone who does the same thing over and over again, expecting a different result.

The first time I saw Ann, I told her the whole sad story of Evan. She calmly listened and then she said, "Jen, was there any point where you thought to yourself: Oh, this is not going to end well?" That question sank into me, and I had to admit there was. I think early on in our relationship, something inside me knew this was doomed. But instead of really experiencing what *I felt* in the relationship—which might have tipped me off that I wasn't nearly as happy as I thought I was—I poured myself into making myself worthy of Evan's love. I was looking for my goodie bag at the end, all wrapped up in a pretty bow.

Ann helped me see how I kept upping the ante in my life. I had to constantly make life more difficult for myself. I could never take the easy road if there was a hard one. In fact, the easy road felt like cheating to me. Anything truly worth having must be difficult, right? Just as I'd thrown my life into chaos around the trial, and I'd constantly raised the bar by which I judged my self-worth, now I'd spent three years with a man I thought I was completely in love with, who'd never given me the slightest proof that he wanted to share a life with me. I started to ask some important questions.

What did love feel like for me?

What did I think love was supposed to be?

I realized that I was confused about love itself. I didn't know what I wanted, but I knew what I didn't want. So until I could figure it out, I went to the one place that always healed me: my business. It had given me a new life after my attack and was there to give me shelter again. For the next two years, I did nothing but work and go to therapy.

Over the years the economy improved, and we rose right along with it. My business shifted into two divisions. One was the free service we provided while being supported by our vendors. The other was back to a traditional consulting model, as half our clients wanted us running and organizing all aspects of their events, not just finding the vendors. Meanwhile, I'd been hearing from clients who did business in other cities that they were unhappy with their local planners. The tipping point was when a major Internet client of mine in Silicon Valley said, "What will it take for you to open a Save the Date® out here?" With one client under my belt there, I thought, If you build it, they will come. I sat down one by one with my sixty largest clients and asked them, "If Save the Date® was in another city, would you use us? And in which cities do you need support?" I proceeded to expand my business to a total of five cities within eighteen months. My staff members regularly flew to San Francisco and Los Angeles to produce our events out there with partnering agencies. At the same time I hired two MBAs, and we traveled back and forth between New York, D.C., and Chicago, rolling out the business in those locations, where I handpicked and hired all the local staff to form new management teams. I did $30 million in business that year.

Then came September 11, 2001, and the world stopped. We were all in mourning, and the idea of celebrating became shameful. In the next few months, economic disaster followed; the Internet bubble burst soon thereafter, and even if there had been a desire to celebrate, few companies could afford the expense. My fledgling businesses around the country withered to nothing, and I was at risk of losing New York as well if I didn't do something—fast. The choice I had was clear. I could hang on to my dreams and expectations and lose Save the Date® altogether, or I could face reality.

So in the course of forty-eight hours, I traveled solo to each satellite office, laid off all the staff I had so carefully hired, shut down the computers, and turned the locks on the doors. It was the most difficult thing I've ever done. I felt like an emotionless zombie—I probably seemed so cold and heartless to these dedicated women, but I could not let myself get emotional. *It's business, not personal*, I kept saying to myself.

When I got home I pulled the shades, lay down on my bed, and sobbed for two days straight. Every time I progressed forward in my life, it seemed, I always suffered a loss.

. . . .

During all the time that I thought about nothing but work, Bennett kept up his gallant efforts to win me over. But to me, he was just being my friend. If he wasn't bringing dinner over to my office, then he was helping me with spreadsheets for my company or looking over my sales models.

Everyone loved Bennett—my parents, my sisters, my friends. On a weekly basis, invariably one of them would look at me and say, "What about Bennett?" And I'd always say no, never. It would be like kissing a brother, I said.

It's not that Bennett wasn't attractive—he was. But all the reasons that I loved him also frightened me. He wasn't at all afraid of me, and he saw through all my masks and bravado. I could be totally vulnerable with him, and that felt wonderful in a friend—but scary in a boyfriend. I'd never had a relationship like that before, and I knew that if anything romantic happened between us, there were only two possible results: either it would ruin our friendship, or we'd end up getting married. And I wasn't ready for either with him.

Despite my discouragement, every week Bennett sent me something—lilies, orchids, a topiary bear, presents, perfume. I told him not to, but it made no difference to him. Once I was excited for him about a new girlfriend he was seeing, and he shrugged and said, "She's just a placeholder." I couldn't help laughing. I said, "Bennett, not for nothing, but you might want to play a little harder to get, because this approach doesn't seem to be working."

But that was the thing about Bennett—he didn't play. He knew I wasn't ready, so he gave me my space. But he never let me forget that he was there waiting for me, and he never gave me one reason to doubt that he always would be. He wrote me the most beautiful, openhearted love letters. He told me that he loved everything about me—and he noticed everything about me. And each letter he signed the same way: "I love you millions, billions, trillions."

. . . .

Sometimes the heart leads and the head follows. In my case, my head and my heart were in competition for which should take the lead. When I felt ready to try a romantic relationship again, I reached out oh so tentatively, still afraid that I'd make the same mistakes. But nervous as I was, I knew I didn't want to be alone forever. At some point, I was going to have to take another chance.

Years before Evan, I'd dated a really tall, handsome man named Charles. He was an associate for a big accounting firm, incredibly smart, and I remember how it blew my mind that he could do calculations in his head faster than a computer. Back then he seemed a little young, a little immature, and after a few months we had nothing to talk about, so I broke it off. Fast-forward years later, we both were single, people mature, so I thought, Why not?

Charles was even more attractive than I remembered him, and he had the same whip-smart calculator brain, had been promoted to Partner, but after about two months of dating we were right back where we'd been seven years before. The man had exactly sixty days of conversation in him, no more. Beyond that, it was an endless loop.

Around this time, I went out for drinks with my good friend Laurie. She was among the very few who knew everything about my past. She also knew everything about my history with Evan, and Bennett's unrequited love. That night she asked me how things were going with Charles, and I audibly sighed. I told her that I was coming to the unhappy realization that my version of love was a fantasy. There was no such thing as the perfect man, and if I kept looking for him, then I'd be alone forever.

Laurie put her hand on mine, and she said, "Jenny, I love you, you are one of my best friends, but you're blind. You have the perfect man in your life already. He's right in front of you."

I looked at her blankly. I said, "Who? Charles?"

If Laurie had been the type to smack someone in the head, she probably would have. She said, "Bennett, stupid. He's the perfect man for you, and you just can't see it."

That night, I laughed at Laurie's suggestion, the same way I always did when someone told me that Bennett was my future. But the words sat with me, and they planted a little seed.

Two days later, it was my birthday, and Bennett offered to throw me a party. He invited some close friends, as well as Charles, of course. Bennett's lovely new girlfriend Isabella would also be there.

The day before the dinner party, Bennett and I went out to lunch at a local diner, and he told me he wanted to give me my birthday present. He handed me a large black box with a lattice weave. Inside that box was another box, and on and on, each box getting progressively smaller until finally inside there was a jewelry box containing a beautiful pair of earrings—silver and gold with bright blue stones. I was shocked and blown away, and of course it was totally inappropriate. He wasn't my boyfriend, and I absolutely shouldn't accept them. But I did, because it was Bennett, and it made him happy to make me happy.

The next night, everything about Bennett's dinner party for me was perfect. It was the kind of evening that could only have been orchestrated by someone who knew me even better than I knew myself. He had all my favorite flowers—vases of lilies and bowls of floating gardenias. He knew I never touched a simple carb, so for dinner there was shrimp cocktail and veal chops. He knew that my no-carbs rule did not extend to alcohol, so he had all my favorite wines. And because he knew that I adored chocolate (no-carbs rule be damned), dessert was my favorite triple-chocolate cake from Black Hound Bakery.

Bennett had designed every detail of the night with me in mind, and I knew why. It was because he still loved me, and he always would. And there he was with his beautiful girlfriend, and here I was with my handsome boyfriend. I can't know what Isabella thought, but Charles didn't even raise an eyebrow. It didn't occur to him—not in a million years—that Bennett could be a threat.

A day after Bennett's exquisite dinner party, I planned to spend the evening of my actual birthday with Charles. By every possible standard, the evening turned out really, really badly. But in order to appreciate just *how* badly it went, it's important to understand a few fundamental things that my nearest and dearest all know about me:

1. I don't collect stuff; clutter makes me nervous. All surfaces in my house are clean. I was a slob before the attack, but ever since I've been a neat freak.
2. I'm a purist when it comes to chocolate. I never mix it with fruit—I don't even like chocolate-covered strawberries.
3. Scents are very important to me; they connect me with feelings and memories —especially after having lost my sense of smell for so long. I love perfume, and candles, and cologne on men.

And here's something it's important to know about Charles: he was born without a sense of smell.

On the night of my birthday, Charles arrived at my apartment with three wrapped presents:

1. A Lladro figurine of a woman holding a birthday cake. Charles's mother collects Lladro figurines. I do not.
2. A box of chocolate-covered cherries. The combination of fruit and chocolate makes me gag.
3. A bottle of perfume. This might have been a lovely gesture from a man who couldn't smell it himself, but not only hadn't he checked what perfumes I liked, but it turned out to be his mother's favorite scent.

I don't know what kind of a happy face I conjured as I opened each of those off-target gifts, but it was convincing enough to carry us to the restaurant where Charles took me to dinner. Then it hit me: this man is gorgeous, but we have nothing—absolutely nothing—in common. Of course he meant well—but he just didn't see me at all.

Then, over dinner, Charles said something that I couldn't forgive, and I couldn't explain away. With a combination of derision and mystification, he said, "What's Isabella doing with Bennett?" I knew Charles through and through, so I also knew right away what he meant, and it wasn't kind. He wasn't saying, *What's Isabella doing with Bennett when he's so obviously in love with you?* He was saying, *What's that gorgeous woman doing with that short man with the quirky wardrobe?*

I set my fork down on my plate and just stared at Charles in disbelief. This man actually had the nerve to feel superior to Bennett—my dearest friend Bennett, who was the truest, best man I'd ever known.

That's when the argument started. I'm sure Charles didn't know what hit him, and he was still in the dark when we lay down next to each other in bed that night. As I heard his breathing deepen into sleep, I thought about how I was doing exactly what I'd been afraid of—I'd chosen a man who looked good on paper and not the man who felt right in my heart. If I didn't stop trudging this same well-worn track, then the walls of my trench were going to be so far over my head that I'd never get out.

I couldn't even wait until morning to break up with Charles. I woke him in the middle of the night and told him it wasn't working and he had to leave, and I meant right that instant. I packed up the three presents he'd given me, and I sent him on his way. He didn't put up much of a fight, and I'm sure he thought I was out of my mind. But I wasn't—for the first time in a long time, I was feeling quite clear and sane. I thought about all the times I'd tried to make someone fit into a perfect box for me, or how I had tried to make myself what another person wanted so that he would love me.

All those years with Evan had led me to Bennett, and thankfully Charles helped me realize it. I finally understood that love wasn't about me trying to be perfect for someone else; it was about understanding who felt perfect for me.

Finally I understood that the kind of love I had experienced didn't make me feel anything but tired and unfulfilled, leaving me to search for the next challenge to prove myself again. Real love is consistent and always there, no matter what, and I was ready to try and feel it for myself. The only man who had ever felt like that to me was Bennett, and why would I ever walk away from that much love?

The next day I went to Florida with my family, and I called Bennett to tell him that I'd broken up with Charles. Then I said, "You know, I've been thinking—"

Bennett replied, "I know."

To this day Bennett insists that he had the whole thing planned out. He'd carefully designed everything about that birthday dinner party to show me that no one would ever know me better than he did. Especially not a man like Charles.

Bennett is many things, but slow on the uptake is not one of them. By the time I got back to New York for New Year's Eve, he'd broken up with Isabella. She was a beautiful girl, and I didn't doubt that she deserved him far more than I did. But he and I had waited a long, long time.

CHAPTER THIRTEEN

After All These Years

On New Year's Eve, Bennett and I were at a party together, and I knew that the time to act was then or never. If I didn't just go for it—embrace Bennett and all the love he had for me—then I would regret it forever. It was time to stop waiting for the fantasy goodie bag, and to start appreciating what was right here in front of me. So that night, I kissed him for the first time. It was more a "jump him from behind," because I totally surprised him by making the first move. That first kiss was clumsy, and totally embarrassing for me, and very, very surreal for both of us. We pulled back and started laughing. It broke the ice for the real kiss that came next. Let me tell you, it certainly was *not* like kissing a brother. Needless to say, we didn't get out of our bathrobes or leave the house for four days. We were old friends but new lovers. It had only taken us nine years.

From then on, we were inseparable. After all that time, there wasn't any such thing as casual dating for us. We already knew each other too well. Of course his friends were suspicious, and I didn't blame them. For years they'd been telling Bennett to get over me—they actually conducted "Jen interventions" with him during which they told him that he shouldn't see me anymore. And now they were worried that Bennett would be devastated if I changed my mind. They weren't wrong to be concerned. A few months into our relationship I began to get cold feet. I could feel myself shutting down. I was scared, really and truly scared, and all the old fears started creeping in. It was all too easy, it couldn't be right. It was a lot of change for me in a short time, and old habits and patterns die hard. I still didn't know: did I deserve his love, was I even able to keep it now that I'd let myself have it? Bennett was talking about summer shares and where we'd rent, and I started to freak out a little bit. So I did a terrible thing. I said, "Maybe we'll be in different houses this year."

Bennett had been around the block with me too many times not to recognize the warning signal I was giving him loud and clear. His face said it all. When I saw how I'd hurt him, I knew that I didn't want to keep going down that road—it was too familiar, and too sad. Anyone else would have just let me have it, but he didn't. That night he sent me flowers, and I wondered, Why can't I let myself accept his love?

I've always said about Bennett, "No one has ever known me better, and no one has ever loved me more." Because Bennett had flown under my radar as a potential partner, I had allowed him to get closer to me than any boyfriend ever had. Bennett had been unattractive to me for years *because* of his love for me. How could I love someone who actually loved me? I had always stayed in control by withholding the parts of me that felt less protected. But Bennett knew everything about me, and he still loved me. Now I had to choose love and choose the real me over all the masks that I'd been wearing—the Jen suit and the red lipstick and the bicycle pants. I had to open myself up and believe that it would all be okay—I just had to say yes.

On July 1, we went out to dinner for our six-month anniversary. Bennett gave me a pair of lovely turquoise earrings. I gave him a letter I had never planned on showing him. I titled it "Timing Is Everything," and in it I explained why I loved him so much.

And then I gave him a box. I could tell the wheels were turning in Bennett's head. He opened the box and saw this very beautiful watch, and he was thinking, Damn, I only got her earrings. I told him that the watchmaker put a hologram on the back of every watch, and then I said: "Let's see what yours is."

He turned it over, and engraved on the back was:

Bennett Will You Marry Me?
MBT

The initials "MBT" stood for *millions, billions, trillions*—his constant refrain in all the love letters he'd sent me over the years. The proposal itself needed no explanation.

No one knew about my plan to ask Bennett to marry me except for my parents, who were dying with anticipation, and were praying I was as confident as I seemed about doing it this way. And of course, the people who'd engraved the watch also knew. They'd all asked me, "Are you sure he's going to say yes?" Now, watching Bennett stare at the engraving . . . I wasn't so sure at all. He was dead quiet, and I was so nervous that I started to babble. Finally I calmed myself, took a pause, and said, "We don't have to tell anybody if you don't want to marry me. I just wanted you to know that after all these years, you're the one."

He said yes.

I first called Bennett's mom and said, "I proposed to your son, and he has accepted." If her complete shock was any indication of what was to come, we had to tell people ourselves, so we could enjoy their reactions. We called the rest of our family and best friends as soon as we left the restaurant to tell them our news, and they were all floored and thrilled. By the time the weekend rolled around, we were sitting on Sagg Main Beach, enjoying the sun, when we heard a couple gossiping on the adjacent beach blanket. "Oh my God," the girl said to the man, "did you hear about the girl who proposed to her boyfriend on the back of a watch?"

We laughed and then leaned over and showed them Bennett's watch. Then we laughed some more, sitting there on the same beach where I'd worn my bike shorts to cover my scars. The same beach where I had spent a lot of time growing up. The same beach where Bennett had spent a lot of time waiting for me. Now here I was, in my bikini for the first time in a decade, and I felt my life was starting again.

I hadn't been entirely truthful all those years that I said I didn't care about an engagement ring. There was one ring that had been haunting me for a long, long time.

I was single when I first saw it in a shop window while I was walking through Manhattan's diamond district. I'm sure I was on my way to an appointment, walking at a fast clip, but that ring made me stop in my tracks. As a wedding planner, I'd seen a million rings. The first thing you do when you meet a bride-to-be is ooh and ah over her ring. But this ring was unlike anything I'd ever seen. The man in the store saw me staring, and he waved to me to come in. I was mortified—I couldn't imagine anything more embarrassing and so *not me* than ogling an engagement ring when I didn't even have a boyfriend. I ran away.

The next time I walked by that storefront, I stopped again. The same man smiled at me and waved me in, and this time I couldn't say no . . . I even tried it on, God help me. He said, "Who's the lucky man?" I told him there wasn't one, and that unfortunately this was the one piece of jewelry that I couldn't buy for myself.

All the while I was single or dating, I had this ring in my head. I remember I'd even draw pictures of it for people—including Bennett. I stayed away from it while I was with Evan—I have a strong superstitious streak, and it seemed like tempting fate. Just once I went by the jewelry shop during my years with Evan, and said to the smiling jeweler, "I finally found the 'one.' Phew!" And then Evan and I broke up.

Since I was the one to propose to Bennett, he wasn't ready with an engagement ring for me. Then a few weeks after I proposed, I kind of looked at him sideways and said, "So does this mean I don't get a ring?" (Very smooth, Jen.)

Bennett said of course I'd get a ring, but there was no way he was going to buy it without my involvement. Then he said, "What about *the ring*?"

We went to the little shop on Forty-seventh Street, and finally I had a real, live fiancé by my side. We walked in together, and I smiled at the jeweler and said, "Hi, do you remember me?" And he said of course he remembered. Then he pulled out my ring, and told me that he'd actually sold it to someone else, but the wedding never happened. The couple broke up and sent the ring back, and it had been sitting there waiting for me ever since. Bennett liked the ring very much, but then he asked me if I wanted a bigger stone, or a different setting, or the same setting with a different stone. All of a sudden I could really have it—*the ring*—and I panicked. Maybe I should shop around, I thought. Maybe there were other rings in the world.

So we looked everywhere, and we took our time, until finally I was positive—there was no other ring in the world for me.

Bennett called to get the ring, and the jeweler told him that it had been sent out on consignment. A client wanted a few different rings to consider, and my ring was one of them. I burst into tears, and there was no calming me down. I finally understood all those poor brides I'd worked with over the years who'd freaked out over the little things. For them it wasn't just "a dress" or "an invitation"—that piece of satin or slip of paper was the symbol of their hopes and their dreams. The same way this ring had become the symbol of mine. I begged the jeweler to get my ring back. And he did.

To this day if you walk into that little shop on Forty-seventh Street they will tell you the story of the crazy lady who haunted her ring for years and years until she finally met her husband.

Pity the event planner who has to plan her own wedding. Bennett had been in thirty-seven weddings already, and his one request was that he didn't want a big, splashy event. I was fine with that. My sister'd had the huge University Club wedding with the centerpieces and the brides-maids, the big band, the hundreds of guests. It was lovely and perfect in every way, but one wedding like that felt like enough for any family.

Bennett suggested eloping, but I didn't feel I could do that to my parents or my girlfriends, who'd be so sad not to see me get married. My parents planned to throw us an engagement party, so then Bennett came up with the idea of having a surprise wedding in the middle of the party. Now that idea had possibilities. But ever the planner, I immediately saw the inherent problems: what if people showed up late, thinking it was just a party? But out of that idea came the one that finally gelled: we'd have a small surprise wedding, but it would be planned to the hilt.

We decided to get married in the fall, and since December is my chaotic season at work, we did everything backwards and went on a honeymoon before the wedding. I took Bennett to my favorite places—Rome and the Amalfi coast. Before we left, all the plans for the wedding were in place, except for one minor detail: I couldn't make a decision about my dress. Physician, heal thyself. Cobbler, put some shoes on your kids' feet.

We were in Rome, walking down the Spanish Steps, when I looked up and saw a bridal atelier. It seemed like a sign from the gods, so I buzzed the door. A musical voice answered the intercom, and I said, "Hi, are you a wedding shop?" She said they were couture and by appointment only. So I babbled into the intercom all about how I was from America, and my wedding was just weeks away, and I didn't have a dress. Thankfully, she found the whole story charming and told me that if I fit into any of her sample sizes, she'd sell me the dress. And I found the perfect one—a halter style with a straightish skirt and lace-up bodice. There was a little train with thousands of buttons down the back. But the most amazing part was the hundreds of tiny raffia roses threaded throughout the lace. It was gorgeous, and it fit me like a dream. She sold me the dress, and I dragged it with me all around the Amalfi coast for the next two weeks.

All my friends knew that the holidays were my busiest time, so when they asked about the wedding date, I just said that I was way too stressed to think about it just yet. In the meantime, we told them that we'd be throwing an intimate little party to celebrate our engagement with our nearest and dearest. Bennett's sister Ellen sent out the invitations, and even she wasn't in on the secret. The invitation read:

> *How do you celebrate the engagement of two epicureans? By having a movable feast, of course. Please join me for a night of interesting locations, great friends, and food. Don't be late, or you'll miss the bus.*

The word went out that the party would move from restaurant to restaurant around the city, and that it would be a fantastic night—not to be missed. We scheduled it the Thursday before Thanksgiving, since we knew that friends and family could tag a few more days onto their holiday travel. The only other people who knew about the wedding were my staff, who helped me plan it, and my sisters and Bennett's sister Ellen, who we also told just a few weeks before because we needed them to participate. Finally, just a few days before, we told my parents.

It would be nice to continue this story on to the happy ending without any more detours, but that's not how life goes.

The night before our wedding, I woke up in a panic—a full-on heart-racing, cold-sweat panic. I wasn't sure I could go through with the wedding. I had worked my way through so many issues, I *thought*, but knowing I was really promising myself to Bennett, in *marriage*, I doubted myself again. Could this happiness last? Was it actually possible for me? My old feelings of unworthiness surged up—along with my fears that nothing good comes without a steep price to pay.

I was alone, and terrified, and there was no one I could talk to. If this had been a typical wedding, I would have phoned up one of my best friends, and we would have talked it through. But this was a surprise wedding. There was only one person on earth I could confide in, and he was the one person who would be most devastated by what I had to say.

I called Bennett, and I told him that I loved him, and I knew in my heart that he was the best man for me, and I assumed I was mature enough to really believe it in my head, but what if I was wrong? Because right now my head was screaming: *Run.* I was so used to being with a man who hadn't been able to give himself to me that Bennett's whole-hearted, unselfish love for me still seemed impossible.

To this day I don't know how he maintained his composure, but that's the kind of person Bennett is. He's the least threatened, least jealous, least insecure person on the face of the planet. He just listened to me like a friend, and he talked me down off that ledge. Then he asked me if I trusted him.

I told him I did.

Bennett said, "Do you trust me enough to show up tomorrow?"

I trusted him more than I trusted myself. "I do," I said.

"Okay," Bennett said, "then just say those two words tomorrow."

When I woke up the next morning, I was calm. The clouds of dread that had been hanging over me in the night had lifted, and I could breathe and think clearly. I went for a run, and I got my hair done. I nonchalantly told the stylist that I was getting married that day, and she almost dropped the hair dryer. Later, when my sisters came over to help me dress, I was happy and in the moment.

I hadn't undergone a miracle transformation in the night, and there were no prescription drugs involved. It was love that got me through. Even if I didn't have faith in myself, I had faith in Bennett, and he didn't let me push him away. I knew that I was exactly where I was supposed to be.

We started at the Villard Bar, a cozy red-velvet-upholstered bar in the Palace Hotel. I wore a vintage Pucci dress, and Bennett walked over with his friends from Bear Stearns. After an hour bells rang, and all sixty guests emerged onto the street to find two shuttle buses to take them to the next stop. Bennett got on one bus, and everyone on that bus assumed that I'd gotten on the other bus. Really, I'd slipped out five minutes early with Rachel and we'd zipped over to a restaurant, Lutèce, right down the street on Fiftieth Street.

Inside Lutèce there was a tiny bar up front, then stairs up to a long hallway that led to the back dining room. So for all intents and purposes when you entered you couldn't see where you were going. When the buses arrived, we set up a diversion with the coat check so that Bennett could scoot out first, and head up the back stairs with just enough time to change his tie. When the guests walked single file up the stairs and down the hall, they spilled into a room filled with red roses, with an aisle of rose petals right down the center. There was candlelight and a chuppah made from my grandfather's tallith.

Each person who entered the room screamed, "Oh my God, they're getting married!" But no one else behind them could hear, so it was a stream of shock and joyful amazement repeated sixty times. Bennett was standing to the right of the rabbi, and my sisters handed each person a candle, which Ellen then lit. Once everyone was up the stairs and in place, I walked in, wearing my wedding dress. While a band played Van Morrison's "Someone Like You," I grabbed my father by the arm on my way up the aisle.

Bennett and I had written letters to each other, and the rabbi read them aloud. They were sweet and tender, but each of us had included a joke at the end. His was that he promised to finally start wearing sunscreen and to keep off the pounds he'd lost. In mine, I told him that he'd finally gotten what he'd been asking for all these years (me). And since he knew me better than anyone, and he actually asked for it, he'd never be able to complain about me. That definitely got a laugh.

After the ceremony the band played "At Last." Bennett had finally got his girl.

We'd rented out Lutèce for a five-course dinner paired with wines, and each person was given an orchid in a color that matched the orchid on the table where they'd be sitting. The first course was foie gras with chocolate sauce—an homage to Bennett's dessert-first philosophy—and my additional surprise to Bennett was that our wedding cake was constructed to look like a gorgeous orchid plant. We had an open mike in case anyone wanted to say something, and by the end of the evening there were dozens of speeches.

The following weekend, our tiny secret wedding was featured as the huge profile in the "Vows" column in the *New York Times* Style section. Can you blame me? Once a wedding planner, always a wedding planner. And it *was* a fabulous wedding.

Bennett had seen me at my worst, and he'd stuck around anyway. For years I'd said no, nope, never going to happen. I struggled and I resisted, but love won out over doubt and fear. It took nine years, but in the end, he was the one. Millions, billions, trillions.

CHAPTER FOURTEEN

Something Is Happening

The first year of our marriage was a rocky road. I still couldn't quite believe that this wonderful new life I had wasn't just a temporary thing. Maybe, I told myself, Bennett was just in love with an idea of me? Maybe now that I was living with him day in and day out, he'd see the real me—and then buyer's remorse would kick in? I kept throwing my old feelings of unworthiness in the way of our happiness, constantly testing his love before I could believe that it was real and permanent.

Just as my commitment anxieties didn't magically disappear once I kissed Bennett the first time, my eating issues didn't simply dissolve once I was in a stable relationship. Bennett loved good food and great wine, and he knew how much I loved them, too. So it made him sad that I just couldn't let myself indulge those desires. While he ate his chocolate soufflés, I kept eating my usual five foods. I had proven myself worthy in so many spheres of my life, and shed so many of my layers, but my food denial was my last security blanket, and I held on to it for dear life.

Bennett knew that I relaxed my eating regimen on my birthday, so he'd say, "Let's just pretend today's your birthday." But even when I did give in and eat something that I loved, then I'd berate myself the next day. I'd do penance by exercising to exhaustion. I'd look in the mirror and say the nastiest things to myself. If Bennett overheard me, he'd say, "Hey, don't say that stuff about my wife." He'd cringe when I looked at my fit body and called myself fat. In frustration and sadness, he'd say, "Doesn't my opinion of you count?" But telling me to stop hating myself was like telling a chain smoker to stop smoking. If only it were that easy.

For six years I didn't get my period, I had very little body fat, and I was proud of it. I thought menstruating was girly and inconvenient. I'd done a lot of bleeding in my life already, and it felt powerful to be able to just stop it. But during all those years of withholding food and exercising my body into submission, it never once occurred to me that all of my self-control would culminate in the worst kind of helplessness.

Prior to the attack I'd wanted a big family—after the attack, of course, that dream got put on the farthest back burner. But marrying Bennett meant that I could actually make that dream come true. The problem was that I harbored irrational, illogical fears about having a child. I knew that I could cope with anything that life threw my way, but how could I be willing to bring children into a world where I *knew* bad things happened? That was a fear that I couldn't just face and shake off—it was way too powerful. And on some level, I think I had to know that if I dieted my period away then I couldn't possibly get pregnant, making my fears a moot point.

Bennett and I tried to get pregnant the natural way for about a year before we went to see a fertility doctor. When I told the doctor that I hadn't had a period in six years, he didn't ask any more questions. He closed the file, looked at me, and said, "Eat a healthy, balanced diet, stop exercising so much, gain some weight, and then come back and see me." I don't like hearing the word no, so I proceeded to shop for a different answer with five other doctors, all of whom said the same exact thing.

Bennett went to each appointment with me knowing full well that the doctors were right, and knowing equally well that I wouldn't allow myself to follow their advice. Every time I ate a regular meal, I'd be seized with panic. To repent, I'd go for a long run the next day and slash my calorie intake. If I tried to slack off on exercise, I turned into my evil twin, taking out my self-loathing on everyone around me—especially Bennett, poor thing.

I so badly wanted to have children, but I couldn't get out of my own way. It broke my heart and Bennett's to see me that way. Finally, in the office of the sixth fertility doctor, I broke down and admitted that I was terrified to gain weight. The doctor looked at me with sympathy, and I knew the solution was up to me. I'd have to trust my husband and relax my suffocating grip on my own body if I was ever going to have the family I wanted.

I made a compromise with myself and with the doctor—I'd begin to eat a normal diet, but I'd also continue to exercise to keep myself feeling strong . . . and sane. Six months and thirteen pounds later, I menstruated for the first time in six years.

Even then, though, Bennett and I couldn't get pregnant. So we began IVF. At that point fertility treatment was a far more embarrassing and taboo subject than it is now. So I didn't tell even our families that we were trying. Meanwhile, every time a friend told me the news that she was pregnant, I'd be excited and happy for her, and then I'd wait until I could find a private place to cry. The concept that someone could get pregnant the good old-fashioned way was almost more than I could bear. I remember wailing to Bennett one time, "Oh my God, people get their kids for *free*!" Those women just had a few glasses of wine, and boom, baby made three. But not me. Once again it felt like the universe was telling me that I didn't deserve the easy way.

The first time I became pregnant, I miscarried at seven weeks. I was devastated, confused, and furious at my own body. I'd done everything I was supposed to do—I'd listened to the doctor, I'd gained weight, I'd taken all the miserable hormone shots and done the IVF. And now it felt even the *hard* road wasn't hard enough for me. I thought, maybe I just didn't deserve a happy life.

It was around this time that my friend Pam invited a bunch of us to a two-layered birthday party. First, she invited all of us who were interested to go to a spinning class with her—she'd booked the entire Upper West Side SoulCycle studio for the occasion. Then, afterward, we were all invited to a lovely meal at a wonderful restaurant.

In the spinning class on the day of the party, the room was lit with candles, and we pumped our legs to all of her favorite music. Something about
the energy in the room got to me. The instructor, Laurie, was incredibly
inspiring, always calling out encouragement, and one thing in particular she
said has stuck with me ever since: *If you feel weak, lean on the group for support—
and when you feel strong, give it back to the group.* Later in the class, when I was
just about out of energy, Laurie called out to me, "Way to go, Jennifer Gilbert!" It was like a shot of positive juju to get me through, and I loved it.

I never could have predicted the degree to which that class would
change my body, and my attitude. I was part of that room, and in that
moment we were all in it together, all showing up for each other. It almost
felt spiritual. I cried in that class over my miscarriage, a loss I hadn't
shared with even the people closest to me because it felt too personal,
too devastating. But days after Pam's birthday, I told my sisters and best
friends about my miscarriage and the grief I felt. It was a revelation to me
that I didn't always have to suffer in silence, that I could receive support.

Since then I have tried to carry the lesson I learned in that spinning
class with me in the rest of my life—when we're strong, we share our
strength, and when we're weak, we find strength in others.

After the miscarriage one of my friends recommended that I see a spiritual
healer who had helped her. By that point I would have eaten toad's poop if
I'd thought it would help me get pregnant, so I made an appointment.

I had no idea what to expect when I walked into Julie's office. Maybe I
expected some sort of swami, but the woman I found was a caring, compas-
sionate Yale graduate. She was a writer who had studied psychology, and was
more of a life and spiritual coach than a magic-wand-wielding psychic. After
days of just talking about my fears and my situation, one day we tried something different. She had me lie down, and then asked me to meditate on
what I wanted. She put her hands over my womb, and she asked me what I
envisioned. In that moment, I told her with complete honesty that the picture in my mind was of a freezer, glistening with frost. She said, "If you were

a baby, would you want to be in there?" How could I hate my body so much—how could I treat it so badly—and yet expect a baby to want to live in there?

How could I work so hard to magnify the light and joy in the world around me, while stifling it in myself?

It was time to make another turn in my life. It was time to stop wishing and hoping that things would change, and to make the decision to change what I could. The rest would have to take care of itself. I couldn't control whether I'd ever get pregnant. But I could warm up that space inside me. I could stop trying to control my body, and learn to love it instead. I could accept my fears for my future children, sit with those feelings, and then release them. I could believe, once and for all, that I was worthy. The only person I'd ever really needed to prove it to was me.

Three months later, Bennett and I geared up to try again, and even before I took the test, I knew that I was pregnant. I could just feel it. And for the first trimester I counted the days and searched for some grace and peace within myself, hoping and praying that this time the pregnancy would take.

After the first trimester, Bennett and I breathed a half sigh of relief, and we told our friends and family that we were pregnant. I bought a few books about pregnancy, and after the first couple of pages I was already in a panic. The title might as well have been *What Jen Should Freak Out about When She's Expecting*. I called my obstetrician, Dr. Steve, to ask him about all these what-ifs. He said, "Jen, I am a high-risk specialist, and I am going to monitor you every step of the way. If you feel anxious, then come in more often. And if those books make you nervous, don't read them. I'll worry plenty for both of us." So I hired a doula he often worked with, and I got ready to surrender control and to trust in the expertise of others.

Working with Julie had sparked a dramatic change in me, and I'd begun to have a new attitude toward food and my body. Pregnancy closed the door on my eating disorder for good. After fifteen years on a low-carb diet, I was eating spaghetti Bolognese five nights a week. For lunch I had man-size subs, and after dinner every night I had an ice cream sundae with chocolate sauce. Bennett, who was so used to begging me to indulge once in a while, would watch me stuff it in and just laugh. Once he looked at me eating ice cream out of the tub and said, "Who are you, and what have you done with my wife?" In restaurants, it was a revelation to me that I could look at the menu and actually pick whatever I was in the mood for. I had denied my food longings for so many years that I had to relearn how to order, how to evaluate my own body's hunger.

Something about carrying a baby loosened me up in a way that nothing else could. I just gave in to my body, and I trusted in its wisdom. Pregnancy allowed me to love my body in a pure, essential way. It wasn't just mine anymore, it was my baby's, too. I even hired a professional photographer and dropped every stitch of my clothing for a complete stranger, without a moment's hesitation. I was bigger than I'd ever been, and I felt beautiful.

While I was pregnant, I had two separate clients who gave me a whole new perspective on my work—and the kind of people that I wanted to work with. I'd always prided myself on spotting the monster clients from a mile away, and just as quickly I'd take off running in the opposite direction. But in both of these cases I ignored my better instincts, and oh, did I pay for it in spades.

Michelle was a bride with every reason to be happy. She was beautiful and wealthy, and her fiancé loved her. But she was proof to me that happiness is elusive for some people. When I first met her, she seemed smart and funny. I caught a whiff of sarcasm in her humor, but I told myself that even if she was spoiled and demanding, that wasn't anything I couldn't handle. Aside from the snark, there was something else that should have tipped me off: she fawned over my engagement ring, but there was no sweetness to her compliments. Her eyes got sharp and calculating, and she said, "I think your ring's bigger than mine." She was newly engaged and should have been in that dreamy phase of being on top of the world, but already she was comparing her piece of the pie with everyone else's.

I took her on anyway, and soon thereafter the crazy train left the station full speed ahead. If I could identify the one thing that sent her over the edge, it was a prominent monthly magazine spread that featured me as an event planner. When Michelle and I visited some spaces the day the magazine hit newsstands, my colleagues made the mistake of congratulating me instead of focusing exclusively on Michelle.

After that, nothing I could do was enough for her, and the strife culminated with her throwing a royal hissy fit because Bennett and I planned to be away for my birthday (December 24, when there's not a whole lot of vendors open anyway), six weeks before her wedding date. At one point she looked at me and said, "You work for me, how dare you go away on a trip?" So I turned to her, and I said, "Lady, you *hired* me, I don't work for you." After that, things got so acrimonious that she actually banned me from her wedding and insisted that someone else in the office handle it. I'd like to say that I just shook it off and said good riddance, but I was horrified that I'd been banished from my own client's event, and I would miss the finale I had worked for months to plan.

While Michelle's wedding was still in the early planning stages, I had an even bigger headache to contend with. It was the summer of the Republican National Convention in Manhattan, and the entire city had been booked up for months. We were approached by a lobbyist who wanted to hire us to plan several parties for him. Rick said he was known across the RNC for his after-parties, which ran from 8:00 p.m. to 1:00 a.m. every single night, for five nights in a row, and he wanted us to plan all of them.

This was going to be no easy feat. We couldn't have the parties in a hotel—they were all booked up for dinners and banquets. So the only option was a party space. We must have shown Rick fifteen spaces, and finally the one he loved—the one he simply had to have—was the entire lobby of an unfinished building that occupied a full block in west Chelsea. There was no electricity, no air-conditioning; they hadn't even installed bathrooms yet. The owner swore up and down that the space would be ready in time, but insisted on a contract that gave him no responsibility whatsoever if it wasn't. Rick wasn't worried, though, and he had his heart set on that space. We negotiated the space fee down and made a verbal agreement. But when we presented Rick with the contract, suddenly he said no. He wanted the space for even less.

This should have been my signal to say, Forget it. But at a time when I should have been reveling in my pregnancy, maybe giving myself a little down time to enjoy something I had worked so hard for, I went against my intuition and plowed ahead with this client.

The plan was to create a miniature city within the unfinished space—Little Italy, Chinatown, the East Village, Harlem, and so on. And all this had to be erected from scratch in this enormous tunnel. There would also be a live band and five hours of open bar every night for five nights.

Two weeks out from the event, we found out that the owner hadn't received his certificate of occupancy, which meant that we were not legally allowed to have anyone in this space. Normally I don't like to bring my clients a problem without a solution, but this was too big a problem not to tell Rick about. When he got the news, he screamed at me so loudly that I had to hold the phone a foot away from my head. When I tried to explain to him what our options were, he hung up on me midsentence.

Of course we figured out a solution and got a temporary license. Problem solved. Except, not really, because the owner of the space then announced that there would be no electricity in time for the parties. He didn't care; he had a contract that said nothing was his fault. Meanwhile, no electricity meant no lights, no air-conditioning (in August), and no working bathrooms. And again, all I got from my client were obscenities screamed into my ear.

The only choice now was to bring in generators—so many that they lined both sides of a city block. For the air-conditioning, we brought in portable units. Instead of producing the party, we became a full contractor to an unfinished space. We were doing things that were not even in the scope of work we had agreed to.

On the day of the first party, we had the generators lined up, the bathrooms working, the air comfortably cool, and the food and music were fantastic. I remember watching Rick walk around the room to inspect everything, and I thought, He's got to be happy. But no, there were no thanks. He walked out of the bathroom, and his only comment to me the entire night was, "I don't like the quality of the toilet paper."

Michelle, the only client who has ever banned me, and Rick, the only client I have ever genuinely wanted to strangle, both taught me something. I would never go so far as to thank them for the wisdom, but I did learn from them: no paycheck is worth that kind of misery. And I promised myself something that has been my mantra ever since: "Outsource everything but your soul." Identify the "soul" of your business (which most of the time is the thing that makes you supremely happy), and hire everyone else to do the rest.

In those two jobs, there was no joy, no satisfaction. And if my soul doesn't feel good at the end of the day, then no amount of referrals, or contracts, or retainers is worth the personal cost. I'd always known that, but I'd allowed myself to ignore it in those two cases. I swore never to do that again.

Mentally those clients exhausted me, but physically I felt great right up to the end of my pregnancy. The doctor told me I'd definitely go late, so I kept on working all day, and then I'd go to the gym to speed-walk on the treadmill. I didn't think anything of it when pain started to climb up my back one evening while I trotted along. After all, labor was in your uterus, right? Everyone knew that.

After working out I went to the movies with Deanna, and when it was over, we walked down the street and I told her that I was having these funky feelings, like something was about to fall out of my vagina. Deanna said, "Okay, we're getting you in a taxi right now." I actually tried to insist on taking the subway home, but she said, "Jen, enough with the tough act." Then she told me that if I didn't get in a taxi, she'd call Bennett. So, fine, I took the taxi. At dinner that night, I kept alternating which butt cheek I sat on, because the pain was still shooting up both sides of my back.

After dinner I went back to working on the seating arrangements for my event. I did call a girlfriend who'd had a baby and described the pain.

She said it was probably Braxton Hicks. I said, "Who kicks?" I was not kidding—I had absolutely no idea what she was talking about. She told me it was false labor pains. That made sense to me.

It was now Sunday night of Presidents' Day weekend, and a huge snowstorm had rolled into the city. It seemed ridiculous to call my doctor now—my girlfriend had assured me it was just false labor, my doctor had assured me I'd go late, and I probably wouldn't be able to reach him anyway. By this point I was peeing or pooping every hour, and the waves of pain were hitting me every five minutes, then every thirty minutes—there was no regularity. Still I refused to call the doctor. I had a high threshold for pain, I told myself; I'm no hysteric. What would the doctor do, anyway? Most likely he'd just be annoyed that I'd disturbed his weekend, and then he'd tell me I was in false labor and to go back to bed.

So I called my girlfriend again. She asked me if I was timing the pain, and I told her that it was all over the place. She said I should just drink a lot of water. So I did.

Then I was still in huge amounts of pain, and on top of it I had to pee every minute, but it hurt to sit on the toilet.

I called my girlfriend again. She said, Take a bath, it will relax you. I took a bath, and I did feel a little bit better.

Not long after, the pain was just as strong and I was starting to get a little panicked, so my girlfriend suggested I drink a glass of wine because it relaxed the uterus. I remember my own doctor suggesting this to me after my amnio, and considering it was my due date, this baby was cooked. A little wine certainly wouldn't hurt the baby, and it would definitely help to keep me calm. So I drank half a glass of wine.

Now I was waterlogged and slightly buzzed on top of it, which definitely didn't help with the constant peeing. And I was still in pain.

Finally I decided to call the doctor, and of course I got the service. When the on-call doctor called me back, he sounded none too happy. He asked me what my doctor had said the last time he'd examined me, two days before. I told him the doctor had said that my cervix was closed. The on-call doctor said, "Well, then, your cervix is closed." He asked me if the pain was regular, and I told him it wasn't. Then he said, "Look, it's the middle of the night, there's a snowstorm, and I'm not your doctor. You can either go to the ER now, or you can wait until the morning and call your regular doctor."

A normal person would wonder where my husband was during all of this, and the answer is that he was asleep. I had been so sure earlier on in the evening that I was in false labor, and I'd just refused to believe it could be anything else. So, I reasoned, what could Bennett do except pat me on the shoulder and tell me I'd be fine? I couldn't sleep myself, so I'd shooed him off to bed with the assurance that I'd be in soon.

After I hung up with the incredibly mean on-call doctor, I went to the bathroom—again—and out came something that I can only describe as an elephant booger. I rushed in, hysterical, and poked Bennett awake, screaming, "What the hell is this thing that just came out of me?" and at two in the morning we both Googled and turned up the answer (it was the mucus plug, such a lovely name), and found out that it can come out anywhere between ten and two days prior to birth. Okay, so I was going to have a baby in a few days. We knew that already. Bennett said he'd go get the car so it would be downstairs and ready when we went to the doctor's office in the morning.

That reassured me for a little bit, but by the time Bennett got back from his trip to get the car, I was babbling. "Something is happening," I kept saying. "Something is happening."

Finally, at 6:30 a.m., I emerged from the bathroom like a caricature of a woman in labor—hair askew, blotchy face streaked with sweat, bulging eyes, my hands braced on either side of the doorframe. I was in horrible pain coming in waves every two minutes, three minutes, five minutes. "SOMETHING IS HAPPENING," I yelled to Bennett.

At 7:00 a.m. we were bouncing along the snowy Manhattan streets, headed to the hospital. On the way, I called the doula I'd hired. Why hadn't I called her before? Because still, in my mind, I didn't believe that I was in labor. The really insane thing is that I didn't even have my hospital bag with me. I was headed to the hospital *without my bag*—even after the night I'd just had, that's how convinced I was that I wasn't in labor. I got the doula on the phone and said, "Hi, I wanted you to know, just kinda FYI, that I'm on the way to the hospital, and I'm sure I'm fine, I'm just feeling a little nauseated."

The doula said, "What do you mean, you feel nauseated? Have you gone to the bathroom?"

I told her that I'd been doing almost nothing but going to the bathroom all night.

The doula said, "Honey, you're in labor. Did you call Dr. Steve?"

I explained the situation to her, and she asked me if I wanted her to meet me at the hospital. I said, No, she shouldn't do anything like that yet—I'd call her from the hospital.

While Bennett parked the car, I was admitted to trauma. I still remember the nurse who examined me, and how I babbled a steady stream of nonsense about how I was sure it was nothing, and my doctor had told me I'd go late, and on and on. The nurse kept quietly examining me, and then she said, "You're eight-point-five centimeters dilated. You're having this baby now."

I burst into tears. I wasn't crying because I was having my baby. I was crying because it finally hit me that I'd been in labor all freaking night, by myself, with *no drugs*. And on top of it, I was an event planner who had left her beautifully packed hospital bag at home.

By the time the doctor came, I was already bearing down. He said, "Okay, on three I want you to push." I remember I actually asked him to clarify what he wanted me to do—did he mean one, two, *push*, or did he mean one, two, three, *and then push*. Maybe I would have known what to do if I'd read one of those books I'd thrown away, but oh well.

Then it was one, two, three pushes and my baby girl Blaise was born.

I had no bag, no perfectly planned-out wardrobe for the hospital, no stuffed animal for Blaise's bassinette or barrettes for my hair. It was just me, and Bennett, and Blaise—which of course was all we really needed.

PART V
Faith

Blame no one.
Forgive everyone.
Thank someone.

—J.P.G.

CHAPTER FIFTEEN

Class Parent

Marrying Bennett, then becoming pregnant, and then having Blaise—all those things opened me up, layer by layer. Now the little bud inside of me was in full flower. For certain, the first and most important step had been accepting Bennett into my life as a full partner. I had never before been in a relationship where there was so much give-and-take. Initially it was terrifying, but over time I shocked even myself with how naturally I was able to mesh my hopes and desires with his.

I had lived on the Upper East Side for my whole life in New York, and everyone I knew was there. Bennett had moved to the Flatiron area before we married. After we got married, he was ready for a change—he wanted to move farther downtown. I decided that he was as entitled as I was to a fresh start, and if it would make him happy to move downtown, then we would. But I was nervous. We were starting our family, and my sister Rachel, who had already had my one-year-old niece Lila, lived around the corner, as well as every one of my "go-to" girls. Bennett, however, had his heart set on Tribeca.

At this point we were a few years past 9/11, but Tribeca was still only a twinkle of what it is today, and to me it was a triangular-shaped no-man's-land. Moving to Tribeca might as well have been moving to another country. Sure, I'd rented out spaces for weddings and parties in that part of town, but those were in converted warehouses. The first time we went down there to look at apartments, I said to Bennett, "Where are all the *people*?" I could swear I saw tumbleweeds rolling down the streets, it felt so desolate. I'd intellectually agreed to move there, but clearly, in my heart, I had my doubts, and the way I expressed them was by being incapable of making a decision about an apartment. Luckily our Realtor was my friend Tracie—anyone else would have fired me as a client. We almost got to contract on at least three apartments, but each time I would flip out before we signed.

By the time I was eight months pregnant with Blaise, we were living in a sublet because Bennett had sold his apartment. This small one-bedroom rental was not the home I had imagined bringing my baby into, and Bennett was despairing that I'd ever really agree to move.

One evening Bennett and I were invited to a dinner party, but I wasn't feeling well, so I sent him without me. I was in our sublet with my feet up when I got a call from Tracie that she had an apartment she wanted me to see right away. It was for sale by the owner, and it wouldn't last. The second I walked into the apartment, it felt like home. I loved it. I asked the owner what it would take to buy it. His response was, "Full ask." So I said, "Okay, it's a deal," and we shook hands. I hadn't even called Bennett, but somehow I wasn't worried. I got him on the phone at the dinner party and said, "Honey, I just bought an apartment, so maybe you should come see it." He was so relieved that I'd finally made a decision that he didn't even take a pause. He just said, "Well, allrighty then!"

At eight and a half months pregnant, the night before we moved, I panicked. I called my sister Rachel in tears, terrified that I'd made a huge mistake, and would be doomed to live miserably in a wasteland forever. When I got off the phone, only slightly calmer, Bennett would have been well within his rights to be completely fed up with me. But he wasn't, and we made a deal: I would give it twelve months in Tribeca, and if I still hated it, then we could move back to the Upper East Side. No questions asked. I breathed a sigh of relief. That's all I'd needed—an escape hatch. I dried my tears, and we moved into our new apartment the next day.

Once again what got us through a tough situation was that I trusted Bennett, and he trusted me. It might take me a while—and I might torture him a little bit along the way—but with Bennett, I always came around in the end.

When we moved to Tribeca, I knew one person, Tracie. For a social person like me, it wasn't just unnerving, it was traumatic. I wasn't convinced that any of my friends even knew how to get to Tribeca, much less that they would come see us there. It was all irrational fear, but the way I dealt with such panic was to go to my fix-it place. I set about making new friends. Some people might do that one person at a time, but not me. I'd always thrived on knowing all kinds of people and mixing them together. Invariably when I planned a dinner party, it would start out with six guests and end up with fifty. Over the years I had denied myself so many things, but I never denied myself friends; I binged on them.

So even before we moved, I had started asking people I knew if they knew any women living in Tribeca. By the time we got downtown, I'd gathered dozens of names and e-mails. I invited them all to a big party and encouraged them to bring their friends. Finally I had seventy-five women in attendance. Bennett thought it was the funniest thing he'd ever seen. Instead of a "Bring a Man for Jen Party," it was a "Bring a Friend for Jen Party."

For those last few weeks before Blaise was born, I was like a little girl at her first day of kindergarten—a little nervous, and a lot hopeful. I'd see another pregnant woman in a shop or a restaurant, and I'd strike up a conversation (*Oh my God, you're pregnant? I'm pregnant, too. Let's be friends!*). After Blaise was born, my whole life opened up even more. It was incredible to me that I could be hopelessly, unconditionally in love with this tiny stranger. Bennett would just smile when he caught me standing over her crib, looking at her in awe that she existed—and that we had created her. I wanted to be her safe place in this world and always to support her emotionally and physically.

Having my baby changed how I wanted to live my own life. I was like the Grinch in the Dr. Seuss story—my heart grew three sizes that day. I learned that my capacity for love was truly endless, and I got hungry for it. I wanted more, and more, and more.

I joined a new kids' gym that had just opened, and they had an opening-night party for the moms. I didn't know a soul, so I asked three women sitting at a table if I could join them. Jennifer, Michelle, and Daniela became my three new mommy friends, and we've stayed friends to this day. I met my friend Haley when we were both walking strollers down the street. She commented on some part of my stroller that she needed, and asked where I got it. I told her that she could have mine, because I was done with it. She stared at me and said, "But you don't even know me." And I said, "So what, you live a block away, and now I do."

Now I had a whole group of new mom friends, and I decided to form a really casual new mothers' group in my apartment. When word got out, I started getting calls from other women asking if they could "join my class, too." I'd just laugh—me conduct a class? *As if.* I was the woman who didn't even know she was in labor until she'd already started to push. But the more the merrier and our group got bigger and bigger.

Something about the insanity and vulnerability of new motherhood allowed me to relate to other women who were in the same situation, and that in turn made me want to open up to them. While in the past I'd felt that the attack was something to hide except in very select company, I now found myself not wanting to hold it back quite so much. I still wouldn't share my history with just anyone, but for the first time I wanted the new friends that I made to know the real me—all of me.

We'd only been in Tribeca a few months when I decided to move my company's offices just two blocks away. When Bennett heard, he said, "Well, I guess this means we're staying."

Moving my office nearby was just a natural progression. I realized I didn't want to spend an hour commuting each day instead of being with my family. I wanted to be able to run home at lunch and nurse Blaise, or take her to a doctor's appointment. And I wanted all the women in my office to have the same options. I'd always felt strongly about creating a company that I'd actually want to work for, and now I truly had it.

I knew that a flexible schedule was essential if I was going to be the kind of parent that I wanted to be. I never wanted to feel that I couldn't be involved in Blaise's school or life because it was too hard to balance it with work. I never wanted to have to say no to anything that was important to me. So at a time when it might have been advisable for me to take on less, I took on more. I heard another mom complaining that there was only one preschool in Tribeca and not enough space for all the kids who needed it. She said she was working on the idea of starting a new school. I didn't even blink before volunteering to help—this was my new community, my new life, and I was ready to roll up my sleeves and go to work.

Naturally I struggle with guilt, but as my husband says: guilt is a wasted emotion. On the days I feel like I'm failing at everything, I take a deep breath and say, I'm only one human being, and I'm doing the best that I can. I know I'm a better mother/wife/friend because I work. After all, Save the Date® was my first baby, and what parent can desert any of their children? They all have to share your love.

One day when Blaise was about five and she had a day off from school, I was rushing out of the apartment to an important meeting. She looked at me and cried, "Mommy, why are you going to work?" She'd never said anything like that before, and I'm sure mothers everywhere can relate to the little catch in your heart when your child looks at you with her doe eyes and asks such a question.

I could have told Blaise that I was going to work to pay the bills, or to buy the new toys she wanted, but that's not the way I wanted her to think of my work—or hers, someday. So I looked at her, and I said, "Because I *love* it. The same way you go to the playground because you love it." And that's the message I always want her to get from me. I do what I do because I love it, and I'm proud of the environment I've created for other working women as well. In the process, I've struck a balance in my life, and I want the same thing for her.

I'd been working with nonprofit clients for a while because I believed in their causes, but after Blaise was born, I started getting involved in a new way. One year a friend invited me to a charity event for Women for Women. This organization works in war-ravaged countries—Rwanda, Afghanistan, the Congo, among others—and offers local women training and support in the launch of their own businesses. Not only do they help one woman as a result, but they strengthen entire communities. This deeply resonated with me in so many ways, and I felt incredible empathy for these women who had been through terrible traumas in their lives, and who now wanted to build something beautiful and resilient out of the ashes of their past.

I called their offices after the fund-raiser and offered to help. By this point Save the Date® was booking tens of millions of dollars in events, and we had tremendous buying power. I could save them money that they could use on their mission, instead of on their fund-raising events. Planning an event for Women for Women evolved into joining their committee, and the same thing happened when I planned an event for the National Institute of Reproductive Health. Reproductive rights and access to safe health care for women and girls had always been important to me, but when you have your own daughter this issue takes on a whole new meaning.

My life had taken so many turns. I'd gone from an innocent girl, to a broken shell, to a wannabe masked avenger. Then I'd resuscitated my own heart again, and I'd found that it had no limits anymore. I had come through such hard times myself that I couldn't bear to just watch someone else struggle in the dark. Bennett teases me that I automatically say "God bless you" to everybody—it could be the sneezing stranger in a movie theater five rows ahead of me. And I say to Bennett in response,

"Everyone deserves to be blessed." I don't underestimate the importance of even seemingly small gestures. If I'm at a social function and I see a stranger who clearly doesn't know anyone, I'll reach out and say, "Come sit here with me" (forever remembering those new moms that included me at their table). And when I pass a babysitter struggling up the subway steps with a huge stroller, then I help her no matter how tight my suit is or how high my heels. I like to think that I have become part of a larger circle of kindness.

One day I was walking down the street with Blaise in her stroller, and I saw a man on the corner poring over a map, looking this way and that, clearly with no idea where he was. I walked up to him and said, "Can I help you?" In a thick Italian accent he asked me where Chambers Street was, and I gave him directions.

Blaise looked up at me and said, "Mommy, who were you talking to?"

"Just a stranger," I said.

"Why were you talking to a stranger, Mommy?"

I thought about that a second, and then I said, "Because he was lost, and he needed help. And that's what we do, Blaise, we help people who are lost."

I had spent so long being lost myself, but through this whole journey of the last twenty years, every day I've been a little more found.

CHAPTER SIXTEEN

All Clear

I was so grateful for the life that Bennett, Blaise, and I had together, and I was eager to add more children to our family. We hadn't had such an easy time getting pregnant with Blaise, so it didn't surprise us that we'd need to go the IVF route again. Once more I started up with the self-injected hormones—and the hoping and waiting.

I had become a new person over the previous few years, but tragedy is tragedy, and it rocks each of us in our own way. So when I lost my next pregnancy at six months (a boy, I learned afterward)—a point in a pregnancy when your baby has already become so real to you—I was flattened. I would have kept our mourning a secret between Bennett and me, but at six months my belly was out there for the world to see, and there was no way to hide our devastation. The day we found out, I sat down at the computer and wrote an e-mail to everyone we knew. I told them all that I'd lost the pregnancy and was grieving. I said that while I knew they'd want to extend their sympathy, I simply couldn't bear it. *Please don't call, please don't e-mail, I cannot even talk about it.*

I'm sure it shocked some of the people who received it—after all, what's more welcome than sympathy in a time of great pain? But I remembered too well how I reacted to other people's sympathy. After the attack, I had absorbed the pain that other people felt for me, to the point that I could no longer feel my own. I remembered how I felt compelled to write a thank-you note for every card and bouquet of flowers. I remembered how alone I felt when people tried to find a silver lining to my experience. I could imagine the version this time—*At least you have your daughter.* Of course that was true, but I couldn't bear to hear it. Later, people would give me so much advice. They'd tell me I *had* to join a support group, I *had* to let myself have a good cry, I *had* to go to therapy. I knew they meant well, but at this point in my life I thought, No, I don't *have* to do any of those things. All I *have* to do is breathe.

Meanwhile, I couldn't allow myself to collapse—I had a daughter who still needed her mother. But the one thing I could not handle was work. I told my office staff that I needed a week—just a week—to get my head together, and not think about anything other than my family. And this wasn't just any week—we were days away from a celebration for Fox News's ten-year anniversary, a massive event that I'd personally shepherded. We'd been working on it for a full year and a half, and had managed to get permits to clear a block of Forty-eighth Street for the outdoor event. We erected a huge clear-span tent so that guests could look up through the ceiling and see the ticker tape newsreel of the brand-new studio above their heads. We had furniture made and branded pillows, a huge red carpet covered the entire sidewalk, and klieg lights lit up the sky. We'd even had the streetlights and parking meters removed. It was a real New York City happening, with throngs of people watching from behind ropes. Rupert Murdoch was in attendance and scheduled to make a speech, and I was supposed to be there to make sure everything was flawless. The old me wouldn't have dreamed of not being there. But the new me had different priorities. There was no Jen suit in the closet that would have been sufficient to cover up my sadness. I just couldn't go, and the event was flawless without me.

. . . .

It was spring when the doctor gave me the go-ahead to try to get pregnant again. I said to Bennett that we'd give it one more go, and if the IVF didn't take this time, I'd just accept it as a message from the universe and take a break.

All the other times I'd tried to get pregnant since Blaise, I'd done everything by the book and then some—I'd given up exercising, coffee, I'd quit drinking, I'd even done acupuncture. This time I let go and I hoped for the best. I worked as hard as ever, because that's what I loved to do, and I exercised because it made me feel good. And I drank coffee and wine, because, hell, if you're not pregnant then you should be able to drink whatever you want.

When I got pregnant again, the feeling was completely different than it had been with Blaise. The first four months I was so sick that just the thought of certain foods made me queasy. If I even looked at a vegetable, I'd throw up. I was so sick that the doctor prescribed the kind of antinausea medication that chemo patients take. I even had to cancel a trip to Turkey with Deanna and my sisters, because the night before the flight I woke up leaking blood. I was terrified that I was miscarrying, but it turned out that the spasms from throwing up had caused a small rip where the placenta attached to my uterus. The doctor put me on bed rest until it healed.

Over the course of the weeks when I was throwing up like mad, Bennett and I kept it top secret that the reason for all that nausea was that I was carrying not just one but two baby boys. We waited until my fourth month even to tell my parents, because my fear of losing one of those babies was so severe. When we first heard their heartbeats, they were both so strong and present that I knew neither of them was going anywhere. I couldn't wait to tell my parents, especially my father. He'd had three girls, I had a daughter, and my sister had two girls. I knew he loved us all more than anything, but I also knew he would love to have a grandson—and now he was about to have two.

. . . .

Not long after I told my parents that we were having twins, they asked my sisters and me to meet them at the University Club. They called it a "family conference." I knew that my father had been having some health issues, so I suspected they had some news for us, but I never expected to hear that my father had been diagnosed with a rare form of melanoma. He told us that he'd already been to Sloan-Kettering as well as NYU Langone Medical Center, and there were various ways of proceeding, but that no matter what, his chance of survival was only 15 percent.

The whole thing felt surreal. I could hear the words that he was saying, but my brain would not compute them. My father was my emotional touchstone—he'd been my mentor, my trailblazer, my cheerleader. And he was invincible. This couldn't possibly be happening.

The sound of my sisters crying snapped me out of my disbelief, and I went into business mode, rattling off questions about the procedure, how long it would take to know if the treatment was working, and on and on. But I wouldn't cry. I just could not allow myself to go to that dark place, and I wasn't going to let my father go there either.

When I had some private time with my father a few days later, I looked at him and I said, "Dad, if you have an angel of the dark side on one shoulder, battling the angel of the bright side on the other, which one wins? *The one you feed.* You are going to fight this, and you are going to be okay."

I knew that I couldn't control the situation—I'd learned my lesson about control at least a hundred times by now. But I also knew that despair wasn't going to help us. He'd always been the one to tell me, "Yes, you can." He'd made me believe that there was no challenge I couldn't face and beat. Now I needed to believe that for him.

The one thing that my father asked of us was that we not tell anyone about his illness. Not for the first time, I realized how similar my father and I were. After the attack, we'd both suppressed our pain because we each were trying to protect the other. Now I completely understood his desire for silence. Once you start telling people, it becomes all the more real, and it takes on a life of its own. He wanted to contain his illness and deal with it in his own way, without having to manage the worries—and worse, the pity—of everyone he knew.

The months that followed were a balancing act for me. I had a deep, instinctive desire to protect my pregnancy from stress. But stress was all around me. Since the beginning of the year I'd lost a baby, become pregnant with twins, and now my father had been diagnosed with life-threatening cancer. By the time I was eight months pregnant and my belly was so massive I could barely move, I was bending over to tie my father's shoes because he was so weak from radiation. Then, just weeks before I was due, he had an operation to remove what was left of the cancer. I spent the rest of my pregnancy hoping for the best, and praying that my father would be alive to see his grandsons.

Despite the looming worries about my father's illness, my body did its job beautifully while carrying the twins. A C-section was scheduled for thirty-eight weeks, considered full-term for twins, and I felt great right up until the end. The doctor wanted to put me on bed rest, but I never gave him an excuse—I had no swelling, my blood pressure was perfect, and I worked and trotted back and forth between home and my office. It's both poetic and ironic that when I finally relinquished control over my body, my body knew what to do.

When they were born, Saxton was seven pounds and Grey was six pounds, and my father was there in the hospital to hold them, my mother standing by his side. He hadn't yet received the all-clear after his operation, but he was alive.

The first weeks after the boys were born was a blur. Because the easy way never seemed to be my first choice, I decided that the twins would get only breast milk, no formula. So, around the clock, all I did was feed them. I did nothing but nurse and pump, and nurse and pump again.

Then, at twenty-three days old, Saxton developed a fever. Since they didn't know if he had just a cold or something worse, he was quarantined in the hospital for four days and given a spinal tap and a slew of equally awful tests. I slept in a chair next to his bassinette and pumped away. I would send bottles back and forth between home and the hospital to feed Grey. Then I would take the day shift at home, and switch with my baby nurse to care for Grey, and to make sure Blaise knew that she hadn't been forgotten in the process. Those were awful, scary days, and I dragged myself through them, meeting first one urgent need and then the next. Then finally Saxton's test came back and thank goodness it was just a cold and we could bring him home again.

Then it was back to the round-the-clock feeding routine. When the baby nurse brought the boys to me, both of them rooting with hunger at the same time, they were like jaws coming toward me. I'd look down at their dear little heads nursing, and while I loved them with every fiber of my being, I did not love feeling like a cow. It was almost impossible just to keep up with the calories I needed to make all that milk. In the middle of the night after nursing both babies and pumping, I'd stand in the refrigerator stuffing my mouth with the blue-frosted cupcakes my friend Haley had sent me after they were born. I could practically eat in my sleep.

I was barely keeping my head above water, but because I didn't want Blaise to feel neglected with the arrival of these two new babies, I insisted on throwing her a party for her third birthday. And this wasn't just a few kids and a cake. This was a hundred adults and at least seventy-five kids, plus face painters, gym activities, and glitter tattoo artists. There was no other option as far as I was concerned. I wanted life to be perfect for my children—and that meant that I had to be perfect, too.

Then I started to itch. All around my middle, I itched and I itched and I scratched until I bled. The next time I took the boys to the pediatrician, I told her that I had a weird rash and I wondered if she'd take a quick look. So I lifted up my shirt and her face fell. "It's shingles," she said. "It happens sometimes when you get really run-down." Of course. I had shingles, and why not?

One day Bennett came home from work, and I was still in my pajamas, working on the computer. I hadn't showered, my hair was greasy, and the light in my eyes was borderline insane. Bennett looked at me tentatively and said, "Hi, honey, everything okay?"

The floodgates opened, and I wailed, "I look like the devil, and I don't even know who I am anymore!"

Bennett said, "Just stop. Stop breast-feeding."

I stopped bawling. I sniffed. "I'm allowed?"

Bennett sighed. As if anyone was telling me that I had to be supermom—it had been my crazy mission from the start. Once again I had no one to prove my mettle to but myself. "Yes," Bennett said, "you're allowed."

Before I stopped breast-feeding, I let Bennett take one picture of me with both boys, each attached to a boob. It's my proof that, yes, I actually did do that. My body, which I'd punished and prodded and whipped into shape all those years, had now carried and nourished three little lives. I promised never to take it for granted again.

After months of waiting, when the babies were just a few weeks old, my father was given the all-clear; he was cancer-free. His treatments had been a success. I had my three kids, my husband who loved me, and my dad was alive and kicking. The universe was looking up.

But if there's one thing I've learned, it's that just because you have one or many bad things happen to you, it's not as if your name is pulled out of the bad karma punch bowl. There's no magical free pass. No, it's always the same raffle of what happens next, and we're all in it together.

Just two months after the twins were born in the spring of 2008, and mere weeks after my father's good news, Bear Stearns—the company where my husband had worked for twenty-five years—collapsed. With it went years and years of savings and a lifetime of career investment.

Now, for the first time in our lives together, it was my husband who needed to be picked up off the floor, and I was the one with enough optimism for both of us. I knew that Bennett would have a lot of grieving to do in the future, but in the meantime I wanted him to know that we'd be just fine. Knowing how hard it was to get out of a bad groove once you started it, there was no way I would let Bennett fall into one of his own.

As Bennett and I lay in our bed with our three children, I thought, My dad is alive, and the people I love most in the world are right here with me. I said to Bennett, "Honey, it's only money."

While we lay there together, the babies squirming and Blaise counting their toes, I remembered back to the night before our wedding. I remembered the simple promise of love and trust that had pulled me through. I learned with marriage and Blaise that love only multiplies, it doesn't divide. So now with the addition of my twins, I simply had more love to give. When I needed it most, Bennett had faith in me—and in us—and I had trusted that faith.

Now I had faith in us, too. And there was nothing in the universe that could shake it.

By that fall the financial world was falling apart. The ripple effect was staggering; every company in every industry was laying off, cutting back. Companies were being publicly ridiculed in the press for any type of unnecessary spending, especially on events. Corporate holiday parties, which at one time were worth $8 million a month for my own company, were canceled. Even if my clients had the money, they wouldn't risk the perception of celebrating at a time when so many were losing their jobs. This economic climate was crippling to the hospitality industry especially. Three catering companies I had worked with since I started my own firm had closed their doors. Restaurants were vacant, and hotel occupancy was at an all-time low.

It was also the December that I was turning forty. Bennett and my family kept asking me what I wanted to do to celebrate. For weeks I told them all, "Nothing." It just felt like the worst timing.

One day I received a call from the *Wall Street Journal* asking what my feelings were on corporate spending, and how it was affecting my industry. I saw what the ramifications of *not* entertaining looked like, and it was job loss for a whole lot of people. At every one of my events, I would stand in the middle of 1,500 people and beam with pride—not at the gorgeous flowers or the state-of-the-art lighting, but at the sheer commerce of the event. Over the years, I had directly or indirectly employed thousands of people: catering directors, waiters, dishwashers, deliverymen, out-of-work actors, entertainers, musicians, taxi and limo drivers. It was big business, and it was helping New Yorkers to support their families, mine included. All of my employees depended on me to make their rent. All of our livelihoods were at stake, and I took that seriously.

With my birthday looming, and given what I had just told the reporter, I realized that doing nothing was setting a bad example. Wasn't my whole career and life's mission built on knowing you *must* enjoy each day?

I decided I would have a small ladies' lunch with just my close friends, at a nearby restaurant or maybe at my apartment. So I thought of caterers. I mean, I am a party planner—why cook when you can cater? Lunch seemed innocuous, so I sat down and started to think of the people who I loved, who inspired me, and had impacted my life. When I counted the names I was at ninety-two women—and that wasn't even including the family members I was inviting—and I couldn't have taken one of them off that list. I started laughing. Well, that certainly wasn't a reservation somewhere. Hell, it wasn't even a private room. With that number of invites, I'd need a whole restaurant. And it surely was not going to be tea and cucumber sandwiches. I called my close friend who managed the Strip House, and since it was never open for lunch, she gave us the whole place to ourselves.

I started to get excited. I needed this celebration after the rough few years we'd had. I wanted a fun, boozy Friday lunch at my favorite steakhouse, with lots of red wine, and great chocolate cake. I laughed about a bunch of ladies digging into creamed spinach and onion rings. Even my invitation was shocking. The front looked innocent enough—it was a long horizontal card with pink, fat, juicy, whimsical letters saying, "Forty &." Then when you opened the twelve-inch card, it said, "fucking fabulous." It was fun and unexpected and powerful, exactly how I felt. In the face of everything I'd lost, and the uncertainty of the future, I wanted laughter and joy and thankfulness for what I did have. I wanted to say, *I love you, and you are all important to me.*

I spent days working on my seating charts and place cards. On one side of the place card was the guest's name and a totally mortifying photo of me (i.e., from the 1980s, a decade that wasn't great for anyone), and the other side had a very concise description of her unique qualities and how we met. Each of those descriptions was different, and not a single adjective was repeated; I wanted these women to know the effort I made to *see* them. When I got up that day to say my toast, I saw my mom and sisters, Rachel and Marissa. Then I looked at all my sisters by choice—the women I had collected through the years, from my first real friend Julie, whom I met when I was seven, to Susie from Camp Fernwood, to friends from Semester at Sea, to all the Tribeca moms who were my new friends. I felt so blessed, it brought me to tears. It was a special, perfect afternoon, more than enough for any momentous birthday. Well, almost . . .

I hate to make anyone feel left out, so the next evening, Saturday night, I had a whopping 250 people over at our apartment for a full-on, DJ-spinning club night. We removed every piece of furniture from the main rooms and lit the entire apartment in purple and red lights. We had glow bars lit up in different colored lights and installed platforms for professional dancers I hired (the former table dancer in me couldn't resist). Every surface was covered in sweets, and at midnight forty boxes of pizza were delivered. It was a total blast. I received over a hundred e-mails and notes afterward thanking me for the best time my guests had had in ages, and for snapping them out of the bad mood they'd been in for months. There have been many, many things I have regretted in my life, but celebrating the end of that year, and the beginning of a new decade was just the *yes, I can* that I'll never regret.

Oh, yeah, and for this party I did give out a goodie bag. It was an empty canvas bag that said "fabulous" on the outside, and it was up to each guest to fill it with his or her own treasures.

CHAPTER SEVENTEEN

Illumination

All any parents want is for their children to be happy. I had experienced that in my own life, and I'd seen it over and over in my clients' lives. The mother of the bride who seems a little too particular (okay, who seems like a complete nightmare) really just wants her baby girl to have the day of her dreams. I couldn't blame her.

I once got a call from a wonderful couple, Debi and Jeff, who wanted me to help plan their dual thirtieth birthday party (their birthdays were within the same week). We had a great connection, and when it was over, they said, "We'll call you for Andrew's bar mitzvah." FYI, Andrew was just three at the time.

True to her word, Debi called me seven years later. Now Andrew was only ten, and the bar mitzvah was still three years away. But for Debi, it was never too early to plan for her boy's big day. She wanted to have the event in the country club where they were members, but she said all of her friends had been there hundreds of times. Rather than choose a different location, she wanted to redesign the country club's banquet room to look like a completely different venue. This would involve custom upholstery and carpeting, and the wallpaper would be Andy Warhol–esque murals of her son's face. Debi was so dedicated to making her son's day unique that we used to joke about her coming to work at Save the Date®.

Two days before the bar mitzvah, I got a weeping and wailing phone call from Debi. I was very familiar with the two-days-before-the-event disaster by now. What was it going to be this time? Not enough tables? Wrong color upholstery? No, it was far worse. The country club for which all that upholstery, furniture, wallpaper, and carpeting had been custom designed . . . was *burning to the ground*. In fact, as she wept into the phone, Debi was standing in front of its smoldering ashes with the sirens blaring in the background.

Okay, this was a little worse than the typical event disaster. Debi had been so intent on every detail being a certain way that I was seriously worried that even if we could find another venue, she'd never be happy. Maybe she'd be like one of those binder brides for whom reality could never live up to their massive expectations.

Amazingly, we found an open banquet room in a country club in the next town over. My team lived there for the next two days, and we cut the wallpaper and shoehorned the carpeting so it would all fit. It wasn't going to be perfect, at least not the way we planned it, but it was flawed in a way that only Debi would see. I was worried that this calamity would ruin her whole experience. After all, it had been her project for three years. I knew I'd be able to see her disappointment, even if others couldn't.

This time, I was the one who shouldn't have been so worried. The night of the bar mitzvah, Debi proved to me that she was no binder bride. Yes, she loved planning the party, but she knew the event was a celebration of her son—not of the upholstery or the wallpaper. Debi wept happy tears during her son's bar mitzvah, and then she partied like a rock star at the reception. At the end of the night, she was the last to leave the dance floor, happy, proud, and glowing—and wearing a sweatshirt that read "Andrew's Mom" in big fat letters. Andrew was happy, so Mom was happy, too. It made me appreciate her and her event all the more.

. . . .

My two boys are so different that I always say that if I hadn't witnessed them coming out of me, I'd never believe that they were brothers, much less fraternal twins.

Saxton is the engineer—methodical and careful, he takes his time and figures things out. He's also a team player—on the soccer field he's right in the mix and loving it. If you hand him a basketball, he'll make the basket nine out of ten times, and if you give him a bat, he'll line it up just right and hit the ball. Grey, meanwhile, we call Evel Knievel. He's the kid who doesn't just climb up the ladder to leap into the ball pit—he throws himself into the ball pit backward. If you hand him a soccer ball, he won't kick it, he'll jump over it. The only way I'd feel sure he couldn't hurt himself was if I encased him in bubble wrap from head to toe.

Saxton and Grey don't look particularly alike either, but around their first birthday something occurred that made their appearances even more different. I happened to look down at the top of Grey's head, and I noticed that his cloud of angelic curls was thinning. A few weeks before he'd had a fever and a head-to-toe rash, and the pediatrician said that his hair loss might have resulted from that illness—or from an allergic reaction to the antibiotic he'd been given. Most likely his hair would grow back all on its own, she said, but she referred us to a dermatologist just to be sure.

The dermatologist couldn't give us a firm diagnosis, but listed about six possible causes. She agreed that his hair might grow back. Or it might be alopecia—an immune system disorder that causes hair loss from mild to total. Some children outgrow it, and some don't. She told us there was nothing we could do but wait and see.

A few months later, Grey woke up one morning, and I noticed that there was a fine coat of hair covering his bald spots. By June of that year all his hair had come back. *Phew*, we thought. It must have been a reaction to the antibiotic or the fever, just as the pediatrician had first suggested.

A year went by, and the twins were now two. In Florida, visiting my parents, I looked down at Grey's head one morning and saw a perfectly round, quarter-size bald spot. I sank to my knees and fought back my tears in front of him. Oh, no, I thought. Here we go again. Clearly his hair loss had not been a onetime occurrence.

Back in New York we went to a series of doctors who all said the same thing: alopecia areata. By the third time we heard it, I had stopped crying. We were prescribed foams and creams for the bald patches on his scalp to help stimulate the hair growth. Grey hated it, but he put up with it like a trooper. Eventually he lost 60 percent of the hair on his head, and my friends could no longer assure me that it wasn't noticeable. Now we needed to make sure that it was just hair, and we took him to a pediatric endocrinologist to rule out a whole series of possible (scary) health causes. Blood tests eventually proved that he was completely healthy in every way—but only after we'd been forced to confront the truly terrifying possibility that he wasn't okay.

It was a profound relief that Grey still was just as healthy as Saxton. And because Grey still had a band of hair around the back of his head, when he wore a hat he looked like anyone else. It was winter, so it wasn't so difficult to keep a hat on him, and we made Saxton wear one as well. Sometimes they both protested, but I insisted, and they knew that if they didn't wear their hats, they couldn't go outside.

The first time Grey's hat flew off on a street near our home, I gasped and my skin tingled as if I were having an extreme response to fear. My pulse quickened and my ears closed up. I leaped after the hat and got it back on his head. There wasn't even anyone on the street at the time, but I remember it took me several minutes to calm myself. Insanely, the last time I had felt that way was when Saxton had almost choked on an apple. I reminded myself that this was not a life-and-death situation. It was cosmetic. Still, no matter how many times I reminded myself how lucky we were that Grey was healthy, I couldn't shake a feeling of doom and despair. We had gone from having three perfect kids to dealing with a *situation*.

It was then that I began to backslide. There were days when I struggled to get out of bed. All the lessons I'd learned over the last twenty years flew out the window. I asked myself over and over, How can this possibly be happening to *my* child? Hadn't I learned life's big lesson that there are never guarantees, and ultimately you have no control over the big picture? I also knew that while I could cope with anything that life had in store for me personally, I couldn't stand the thought that one of my children might have to suffer. I could scream at the injustice of it. So I went back to making deals with the universe—I'd give up any amount of pleasure if it meant my son's hair would grow back.

On the days when I felt bleakest—flat on my back in bed, without the slightest will to get up—I was right back in those dark days after the attack, when I'd lost all hope that there was joy in life. The only difference between now and then was that in this situation my sadness wasn't buried under the surface. I wore it right on top of me. If I was at a party and someone asked how I was, I'd burst into tears.

I was so desperate that I began actually to listen to the people who tried to make me feel better. After all these years, I had become an at-leaster.

At least he's not a girl.

At least he's healthy.

At least it's only the hair on his head.

At least he has his eyebrows and eyelashes.

At least he is not in physical pain.

Yes, I thought. That's right. Isn't it? Get a grip, Jen.

In the spring, Grey's hair started growing back again, but he was left with two shiny bald spots. For a year and a half we'd been on a roller coaster of watching, and waiting, and hoping for the best. Each time his hair grew back, we thought, *Phew.* And each time he lost it again, we thought, Oh, no. But still we hoped, and constantly looked for silver linings, because we couldn't entertain the thought of the worst.

In July Grey developed a fever, and the thing that we'd feared all along happened.

By the end of the next week, all of his new hair growth was gone, and the bald patches were getting bigger. One day when I picked him up from his nap, his crib was littered with hair. It was like a mass exodus, and Grey wiped his face to get the fallen strands out of his mouth and his eyes. That visual alone was heart-wrenching.

Eventually Grey lost all of the hair on his head, but he still had eyebrows and eyelashes. With a hat on, he looked like the old Grey.

I was in such denial over what was happening that I refused to take pictures of Grey. It was almost superstitious—as if taking pictures would make it all the more real. I'm embarrassed to say that for months of his toddler life—a time when most parents are snapping pictures every second—we have none.

Around this time, it was my parents' forty-fifth anniversary, and my sisters and I decided to give them the gift of a photo shoot with the whole family. It would be a celebration of so much—my father, cancer-free for four years, my parents' marriage, and their five beautiful grandchildren—and also how far each of their three daughters had come.

But first I had to get over my anxiety about photographing Grey. I had so much to be thankful for, I told myself. I had my three beautiful children and a husband I loved. He'd found his new dream job, and I loved my work. But as I picked out a sweet little sailor hat for Grey to wear in the photograph, I struggled with my sadness.

That afternoon we were on the lawn in the back of my parents' house, and I thought, If this grass could talk. It was the same lawn where I'd thrown my first parties, and where I'd memorized every flower in my mother's garden. It was the same lawn where Rachel and I had played, and baby Marissa had chased after us. Now, on the same lawn, the cousins were giggling and doing cartwheels. Grey's face shone with delight and happiness—an emotion that was captured on all of our faces in the photo that now hangs on my parents' living-room wall. Little did I know that Grey's alopecia would progress, and very soon that photograph would be the only thing I'd have to remind me of what Grey's face looked like before he lost all his eyebrows and eyelashes.

When Grey's last eyelash fell out, I held it in my hand, and I mourned for the loss of that little lash like I mourned my own innocence. I tortured myself with the thought that my son's life would never be the same.

What if the kids at school make fun of him?

What if everywhere he goes, people stare at him?

What if he can't find love?

What if he doesn't get jobs, or girlfriends, or anything else he wants in this life?

I went to my bad dark place. Coincidentally, it had just been the twenty-year anniversary of my attack, and a lot of my own bad memories and pain were bubbling in my head. Now, I told myself, Grey would have his own before and after, and his life would never again be so perfect. And how cruel that he was a twin—I pictured my boys on their birthdays, one with his thick head of hair and the other bald.

With all of Grey's hair now gone, I knew this was alopecia universalis—total hair loss—and I knew that some people's hair never grew back. I pored over Web sites and read devastating stories about adults whose lives were destroyed because they were afraid to leave their own homes. Oh, God, I thought. No. I shut off the computer and told myself never again to Google alopecia.

Every day, no matter where I was—in a work conversation, sitting at dinner with my family—at least 70 percent of my brain was occupied with worry over Grey. I kept imagining those nightmare scenarios in my head. Grey's baldness broke my heart every day, and I couldn't snap myself out of it. None of the "at leasts" that I'd allowed myself to entertain were working anymore. When a well-meaning mother at school said to me, "Well, at least he's a boy," I looked at her in tears. I said, "Oh, really? And what if it was your beautiful boy? Would you take solace in that?"

I became even more of a fanatic about covering Grey's head—indoors and out. Even in warm weather, at preschool where everyone knew him, I insisted he wear a baseball cap. I dreaded people's stares in the playground. I knew that when other parents saw my child's starkly bald head and face they suspected the worst—that he was suffering from cancer. And even while I thanked God that Grey was healthy, I felt a hot rush of shame that these other parents pitied me and my poor child for something that wasn't even true.

Once, on our way home from Jamaica and stuck with plane delays due to weather, my whole family was waiting on line for a connecting flight. It was one in the morning, we had to go through customs, we had a zillion pieces of luggage, and there were hundreds of people on line ahead of us. Basically, we were in hell. Saxton was screaming and clinging to my leg, and Grey was shrieking and rolling around the floor all strung out on Benadryl, while Blaise stood there laughing at the whole mess. I couldn't get Grey to stand up and walk, so I had to keep nudging him forward with my foot, like a piece of luggage. I dreaded the annoyed looks from the other passengers, who must have been totally fed up with my screaming kids. Then I looked around and saw the stares I was getting from other people. They weren't annoyed at him, they were pitying me. They all felt sorry for me, because they thought my kid was sick. Meanwhile, I wanted to say to all of them, "He's not sick! He's just a pain in the butt!"

In more ways than one I was back to my old coping mechanisms—there was the shame and the covering up, and there was also the belief that somehow if I worked hard enough, I could force everything to be all right, as if I could cure my son's alopecia by sheer force of will. Traditional medicine clearly wasn't having any impact on Grey's hair loss, so when a friend mentioned a doctor of Eastern medicine who had helped her through infertility, I didn't see what it could hurt to at least visit him and see what he had to say. My own life had been transformed by a healer, so I was open to anything.

I could have kissed that elderly Asian doctor across the examining table when he told me that he believed that Grey's hair would grow back. He prescribed some teas to brew at home—all perfectly safe, although they made our entire apartment smell like ass—and Grey drank them like a trooper. When the course of teas was completed, the doctor had me apply warm compresses to his body once a day, and that became a ritual for the two of us for weeks and weeks.

I badly wanted to believe that the treatments were working, but there was no evidence. Increasingly, I was snippy with everyone around me, including our babysitter, Pauline, who is like a member of our family. There's no one on earth I trust more, but during this time I was so deeply wounded that I even lashed out at her once because the pantry was a disorganized mess. Finally, after a raised-voice argument, as I was crying and apologizing to her for being a total nightmare, I screamed, "I'm just so mad!"

That moment was very profound for me, and I realized that I had just had a breakthrough. All along I'd been thinking I was sad or deeply upset, but far more so, I was angry—deeply, poisonously angry—a very different emotion. It made me feel clearer, and saner, to be able to identify what I was feeling. Clearly it had nothing to do with Pauline, or the kitchen cabinets. I was in a rage over what was happening to my son, and none of the positive self-talk or warmhearted wisdom that had accumulated in my soul could penetrate my fury. I knew this feeling of betrayal because I had lived with it for a long, long time. And after twenty years it was just as fresh and raw.

I told myself that I was keeping my pain under wraps and that the kids were far too little to pick up on it. Every day I came home from work to see the twins barreling down the hall to greet me. And every day I would have to suppress a gasp at how Grey looked so different from my mental image of him. In my mind he had hair, so the reality of his little face now stung me each time I walked through the door. I'd plaster a smile on my face, then go into my room and weep.

This couldn't go on, and I will be forever grateful to my daughter that she brought an end to it once and for all.

One day over that summer, Blaise was having a play date with a friend. When the little girl caught sight of Grey, she asked why he didn't have any hair. I stiffened, but I didn't say anything, and the moment passed. After the play date was over, I noticed that Blaise seemed sullen, and not at all her usual self. Blaise is a deep-feeling little girl, so I knew something must be bothering her—maybe she and her friend had had an argument. So I asked her what was wrong. She got upset right away, and

I assured her that she could always tell me anything. Then she said, "Mommy, I'm sorry, I forgot to tell Susie the rule."

I looked at Blaise and said, "What rule, honey?" Blaise said, "The rule that we don't talk about hair."

Whoah. I'd never articulated it, but of course she was right. The word *hair* had become verboten in our household. It wasn't spoken of—except in whispers late at night, behind closed doors when the kids were asleep. I told Blaise not to worry, that it was fine, but inside I was rocked. Clearly this facade of normalcy was not fooling her. Then she said, "Mommy, why is Grey's hair falling out?"

My little girl's question—and her frightened face—illuminated the dark circle I'd been walking. Suddenly I understood so much about this path I'd been on for the last twenty years. I understood why my mother had wanted to change the channel when I sat in her living room, bleeding on the sofa. It wasn't because my mother didn't care—it was because she couldn't cope with the fear and the pain, and she wanted to shut it out. My mother didn't know that she was shutting me out along with the pain—she was just doing the best she could with the tools she had. Back then, no one around me wanted to talk about the attack for fear it would upset me. As a result, my attack became something to be ashamed of, to minimize, and to bury. It had taken years to uncover and work on the issues it caused me. Now I was turning Grey's alopecia into a subject of fear and denial as well. Something devastating was happening, and instead of confronting it head-on and recognizing it for what it was—no more, no less—I was not-so-quietly losing my mind over it, and I was dragging everyone else down with me.

I'd allowed myself to forget everything I'd learned over decades and to fall back into those old painful patterns. But by some grace I didn't even know I had, I'd taught my daughter a new pattern. And now she had the inner strength, empathy, and bravery to hold up her hand and point out the elephant in the room. Yes, what was happening to Grey was unknown and scary, but I had forgotten that my children's only reality was *my* reality. And I had to be their center pole.

That minute, I sat down with Blaise, and I called in the boys as well. I looked at all three of them, and I said, "We're going to talk about Grey's hair." I had no plan, no idea whatsoever what I was doing, or preparation with Bennett, but I knew I had to do something immediately. Not for one second longer did I want my home to be a place of silence and shame. So I told them that Grey's little body had an allergy. I knew they could understand that—they had plenty of friends with allergies. I said some people are allergic to bee stings, some to wheat, some to peanuts. We didn't know what Grey's allergy was yet, and his hair might come back or it might not. But in the meantime, he was perfectly healthy, and that's all that mattered.

That conversation was like flipping a switch in my kids' faces. At first they were shocked that I'd said the forbidden word: *hair*. Then their long-buried curiosity took over, and understanding dawned on them like sunshine. *Grey has an allergy—okay, let's go play!* And that was it, they were off and running.

A few weeks later, when Grey and Saxton were about to head off to preschool, I asked Grey if he might like to take his hat off when he was inside (outdoors it protected him from sunburn). He said that he would—and it was like he was admitting something to me that he'd been holding in for a long time. Then he headed off to school and he never looked back. He had never cared about where he was in his hair/no hair cycle. I was the one who needed to let go of my sad predictions and woes. This was his little life, and I needed to let him live it, and if it didn't bother him now, then it was time for me to brighten up my reality.

One night I was watching television, and I saw Bruce Willis—bald as a cue ball and standing there being interviewed on a red carpet alongside his hot girlfriend. I pressed record on the TiVo, and I spent the rest of the night going to every channel looking for more bald men—I scoured basketball games, movies, television shows. By the end of it I must have had thirty different ten-second clips. The next morning I showed Grey one clip after another, starting with Bruce Willis. I said, "Look, Grey, who does he look like?" He looked at Bruce Willis, and then at me. With a huge smile on his face, he said, "Mommy, he's a Grey-Grey man!" We watched each of these men, and I said, "See Grey, some people have hair, and some people don't. Some people have curly hair and some people have straight. Some people have lighter skin and some people have darker skin. Everybody's different, and everybody's beautiful, and it's all okay." Then I handed him the remote control, and I said, "Go get your brother and sister and show them, too."

Everything changed after that. Suddenly, bald was beautiful in our household, and I realized that there was no one more qualified than me to teach my child the gift of resilience and the importance of empathy. I didn't kid myself that Grey would never hear an unkind word, or ever have reason to feel self-conscious. But when that moment came, we'd both have a choice—my choice would be to pick him up and dust him off, and his choice would be to shake it off and keep on loving himself for who he was. We could both choose a brighter path. It was time to open the shutters and let the sun in.

I remembered all the times I'd wished that my own scars were more visible, so people could know just by looking at me that something terrible had happened to me. Meanwhile, here was my child, perfect in every way, and yet he looked like something devastating had happened to him. If I believed in fate, I might think that there was a beautiful kind of kismet in this. I think Grey and I have a lot to teach each other about looking past the surfaces that we all construct for ourselves. I would teach him and his siblings *now* what I had come to understand in my forties: that once you stop fighting yourself and things you can't control, and surrender—just let go—*then* you find your power and the energy to move forward. *The power comes from the surrender, not from the fight.* We can all have a do-over, if we just let ourselves.

We all have our stories—Grey's would just be a little more visible than other people's.

After the night of the TiVo clips, we saw Grey-Grey men everywhere, and Grey loved it. He just glowed with the idea that there were people like him all around. There was nothing wrong with being a Grey-Grey man—in fact it was pretty cool. And then one day Sax announced that he wanted to be a Grey-Grey man, too. For his next haircut, he said, he wanted his head shaved so he could be just like his brother.

So one day, after weeks of pretty constant begging from Saxton, I took all three kids to the barber, and Saxton told the barber he wanted to be a Grey-Grey man. He had all his thick, luxurious hair shaved right off. The squeals of laughter were such a wonderful sound that they're still ringing in my ears. Blaise was cracking up, and Saxton had a smile a mile wide, and he and Grey were kissing and hugging the whole time. It was the most pure and genuine display of love I'd ever seen. I stood there crying, and in that moment I knew that it was all going to be okay.

We plan, and God laughs. My whole life is a lesson in the truth of that statement. I could plan every turn of the road I was on, but I couldn't control my destination any more than I could predict that the bride's heel would break on the way up the steps to the church. Recently my entire office spent nine months managing a concert in Central Park that sixty thousand people would attend. And what does any event planner know about outdoor events? *Hope for sunshine, but plan for rain.*

And of course it rained, the moment we started the concert. It had been clear skies for the entire day, so the event organizers made the call . . . game on. The strange thing was, it only rained in Central Park. I think all the lights and the generators must have caused some kind of localized weather . . . I mean, right? Who could have predicted that? But Andrea Bocelli sang anyway, and it was the amazing capper to months of events that we'd managed all over New York leading up to the concert. I knew the rain wasn't my fault, but in the old days, my surface calm in the face of potential disaster would have been a big fake—underneath I would have been a sea of anxiety. But now, the outer and the inner me are a much closer match.

. . . .

Our neighborhood in Tribeca, which I refer to as Triburbia, is like a little suburb, and now that my kids are six and three and a half, pretty much everyone knows us. Grey knows he has to wear his hat in the sun, but otherwise, his bald head is out there for the world to see. And the people we see on a daily basis are so used to seeing him that they don't even notice it anymore—they just see smiley, gorgeous Grey.

One day at a playground in another neighborhood, I spotted Grey standing at the top of a ladder on a huge jungle gym. He looked at me and he just smiled. Evel Knievel had a plan.

I told him to climb down—*carefully*—but before I could reach him, he jumped. Luckily the ground was covered in thick foam, so when Grey hit planet Earth with a thud, he was shocked and scared but completely fine. Still he was crying his head off, his hat had flown away, and his face was raining tears and snot.

So now it was a scene, and all the other moms were surrounding me and asking if Grey was okay. I knew what they were all thinking—that the poor little cancer kid fell and hurt himself. But the moment passed quickly. Grey stopped crying, and the first intelligible words he said were, "Can I go on the swing?" Meanwhile, I'm thinking, For the love of God, this kid is going to be the death of me.

Grey started walking over to the swings, and I saw this little clump of kids standing there staring at him and whispering. Grey's no dummy, and we'd been in this situation with the bald thing enough times to know what the looks and pointed fingers meant. I hovered close but stayed silent, coiled to jump in at the first sign of trouble, when one little girl finally raised her voice and said to Grey, "You don't have any hair!"

Every atom in my body wanted to say something—do something. But I didn't. My job was to prepare Grey, and give him the tools he needed to live his life to the fullest. So I hesitated just long enough for Grey to respond himself. He looked at that little girl, he smiled like the sun, and he said, "I know. But I'm *cute*." And then he turned around and ran to the swings.

Epilogue

You yourself, as much as anybody else in the
entire universe deserve your love and affection.

—BUDDHA

I once planned a lavish wedding in the Bahamas, and I was responsible for every detail except for one: the cake. The bride had arranged for someone local to make the cake, and she was obsessed, and I mean *obsessed*, with everything about it. She'd been so sweet and easygoing about everything else, but the cake was a different story. That cake became the focus of all her hopes and dreams. It was her marriage, her future, everything she wanted from life, all stacked up in layers of sponge and buttercream.

I didn't fault the bride for sweating that one detail—she was entitled. And I wanted everything to be perfect for her and for her fiancé. They were a lovely couple, and he was an incredibly dear man. He'd worked on the top floor of the Twin Towers, and had been safely out of the office when the planes hit the World Trade Center, but he'd lost so many friends—so many people who would have been there with them to celebrate their big day. Now he just wanted his bride to be happy, and I wanted to give them both the fairy-tale day they deserved.

The wedding was being held at the ultra-expensive Ocean Club at 6:00 p.m. All was in place and going according to plan, but for one tiny problem: the cake never arrived. I'd tried calling the baker, and the club had tried calling her as well, but there was no answer to any of our calls.

Part of me wanted to believe that the baker wasn't answering her phone because she was already on her way to the club to deliver the cake. Another part of me was convinced that she was sitting on a beach somewhere, sipping a cold one and eating my client's wedding cake.

At one point late in the afternoon the bride came up to me and said, "Is the cake here? Have you seen it?" She was aglow with expectation, so I gave her the only possible response: "Yes, I have seen the cake, and you'll die, it's so gorgeous." Meanwhile, I had never even seen a *picture* of the mythical cake—all the bride had told me was that it was chocolate and vanilla, and decorated with seashells.

Maurice, the maître d', was an elegant man of military precision who wore the crispest, whitest uniform I'd ever seen. He seemed utterly unflappable, but when I told him that the cake was missing, I thought I would lose him on the spot. The entire staff knew how insane the bride was about this cake. So Maurice said, "I just have to tell her that there's no cake."

My response? "And ruin her ceremony and the whole day? No way in hell are you telling the bride there's no cake." Then I said, "Maurice, you are going to call your chef and tell him to make me a cake. It's going to be chocolate and vanilla, and it's going to be done in two hours."

Meanwhile, I went to the hotel gift shop and bought ten boxes of Guylian Belgian chocolates shaped like shells and starfish. Then I asked the kitchen to position the chocolates all over the cake, and to paint each one with vanilla buttercream so the chocolates would look like a seamless part of the cake. Then I went around the hotel collecting the silver dollars and gold-painted shells that they used as decoration. These I scattered around the display table where we'd put the cake.

Finally, the moment of truth. It was time for the cake to be rolled into the banquet room before God, the bride and groom, and all their family and friends. Best-case scenario, the bride would take one look at the cake, bite her lip, and have a nervous breakdown later.

Worst-case scenario, she would rend her veil, run screaming from the hotel, and throw herself into the ocean.

At least ten times in the last half hour I'd been on my walkie-talkie with the kitchen, checking on their progress, and I'd done everything I could to delay the unveiling. I don't know who was sweating more, me or Maurice. I'd convinced the bride that the cake had arrived hours before and was chilling in the walk-in refrigerator in the basement. But instead of being rolled in from the elevator, it was perched on the backseat of a golf cart, zooming over from the kitchen, where the last chocolate had just been painted with buttercream.

As the cake slowly made its way to the bride and groom, Maurice and I stood next to each other—hearts racing, hands held, braced for the worst.

The bride took one look at that cake, tears rolling down her face, and she said, "Isn't it beautiful?"

It took me a long time to learn that sometimes in life things just happen, without our permission or planning. Even the most beautiful things can happen that way. Maybe the cake bride loved her wedding cake because it really was perfect, thanks to the minor heart attacks that Maurice and I had endured—but more likely, she loved her cake because she sincerely wanted to be happy. Perception is reality. She was in love and her life felt full; that was her perception, and that became her reality as well. We all have the choice and the power to create the reality that we want. That is a life lesson that I carry with me every day.

I've planned every kind of event the human imagination can dream up—events where every detail is so thought-out that the guests talk about it for weeks (even years) after. Those events are wonderful, and I've loved them and been honored to be a part of realizing other people's joy. But the one downside of being an event planner is that you feel a lot of pressure to carry that kind of over-the-top display into your own life (case in point, my fortieth birthday weekend extravaganza that started out as a ten-girlfriend lunch). Even your own dinner guests have expectations when you plan parties for a living.

For the past few years, we have invited our close friends the Fishers out to our house at the beach for the Fourth of July weekend. The high point of the weekend is usually a seated dinner for fifty, complete with linens, china, an array of hors d'oeuvres, and centerpieces made by yours truly. I line the table with real grass sprouting up flowers, and attach paper butterflies on wires so they appear to be fluttering around the table. I hand-paint vases, hang candles, and put on a full-scale production.

When I called to make plans for this Fourth of July, the first thing my friend Jennifer said to me was, "Jen, I love you, but please no party this year. We just want to be mellow and hang with you."

And that's a true friend. I was so relieved that I could have made out with her through the phone lines.

So instead of linen and crystal, it was paper napkins, three families, and Bennett's grill weighed down with burgers and steaks from Costco. Dessert was a red, white, and blue ice cream cake from Carvel—the comfort food of my childhood.

The twins were ecstatic, running after the older kids like maniacs. Blaise wanted to party-plan the kids' table, so she picked her own flowers and wrote out place cards on the front of seashells. Everyone was cracking up because her table looked lovely—meanwhile I'd slapped some dime-store Fourth of July paper plates on my table and called it a day.

The kids did chalk paintings on the patio and ran around the yard on a scavenger hunt. I ran after them, a sundress thrown over my bathing suit, not a stitch of makeup on my face, and no lipstick in sight. I was doing figure eights all over that yard, and there wasn't another thing on my mind. Joy was all around me, and I was right there in the middle of it. I was no longer the woman who only helped *other* people celebrate life—I was the woman who celebrated it herself, and I'll never stop feeling grateful for that gift.

The scariest step is always the first one. When you're deep down in your own endlessly looping track, it's really hard to see that you're going around in circles. And the way out of that groove can seem like an unbridgeable gap. But sometimes the only difference between being in a hole and getting out is just one footstep. One little boost up, and you're off in a new direction. You may not know where that road will take you, but you know it's away from where you've been standing.

For so many years I believed that the only way I'd ever achieve anything in life was to fight for it. If I fought for my dreams, then maybe I would deserve them. Soon, the fighting became more than just part of the journey—it *was* the journey. After decades of fighting for everything, culminating in my fight to heal Grey, I had an epiphany one day. What if I was wrong about the fight? What if I wasn't marked for disaster? What if, instead, I was actually a blessed soul, and wonderful things would come to me no matter what I did? What if I actually *believed* it was all going to be okay? If you believe in God, maybe you call it faith; if not, maybe you call it fate. Either way, doesn't it make life easier to go through each day knowing that you will be able to handle whatever comes your way? Instead of spending so much energy on worrying, doesn't it make more sense to spend that same energy on creating and enjoying the journey?

Everything in my life has taught me this:

You can't control what may happen to you in this life, but you can control who you want to be after it happens.

It's a very simple, yet powerful statement. Instead of fearing what will happen for my children in the future, I can just love them for who they are right now. Instead of fighting my body, I can give thanks for it. Instead of questioning my husband's love, I can accept it with open arms. And instead of worrying about life and what it has in store for me, I can throw my hands up in the air and enjoy the ride.

Acknowledgments

When I actually sat down to start this book I knew it would be a scary and possibly painful journey, but I had no idea to what levels. While ending this book gives me a sense of closure, I am left with many new internal questions to ponder. I realize more than ever that I am a work in progress and that no matter how much I know, or think I know, what I don't know is far far greater.

Even as of a few years ago, I thought I would never, ever tell (or write) my story, and look at where I am now. Putting my past down on paper makes me feel so exposed, like I am dangling naked from a fishing line over Times Square on New Year's Eve. So, when I walked into HarperCollins for my very first meeting and the entire crew in that room offered such heartfelt encouragement, it literally brought me to tears. Thank you to Jonathan Burnham, Kathy Schneider, and everyone at HarperCollins for bringing my memories to life, and understanding that sometimes the darkness and the light can go hand in hand. Thank you to my editor, Gail Winston, for her insight, perspective, and strong commitment to quality. You have helped me become a writer, not just a reader. And of course, Maya Ziv, for your kind words and wonderful point of view. Tina Andreadis, I'm grateful for your help in promoting this book from beginning to end. Mark Ferguson, thank you for tolerating my endless virtual marketing questions, and to Katherine Beitner, my publicist: I knew at first type I was in the most capable of hands. Thank you all for understanding my need to get this book right.

There are many others to whom I also owe enormous gratitude:

Sam Chapman from Empower Public Relations. Thank you for getting the word out, for all your perseverance, and for being a PR guru as well as family. And to Laura Berman, what can I say, what will I ever be able to say, but *I love ya real good*.

Richard Abate, you were the calm in my planner-detailed, control-freak storm, and I thank you for sharing your time and your thoughts with me, and I do love it when you talk truth to me. Really, really. You were a great agent, but a better friend and advisor. And a huge hug to both Melissas in your life, for their help along the way.

Peternelle van Arsdale. I gave you scrambled eggs, you gave me back a soufflé. You hung in there with me, and I know I'm not easy. Your dedication was incredible, and I could never have done this without you.

The Save the Date® family. Thank you will never be enough. To every employee, current and past: you are the very best at what you do, and I am humbled by the devotion and dedication you feel toward our company. You all have taught me so much, and forced me to be better. To all of our loyal clients, thank you for all your professional and personal support over the last twenty years. And as we are only as good as our vendors and venues, thank you for helping me look good for all these years.

To my incredibly supportive book club (and my mini focus group). You have been living this project with me for over a year now, and you will never know how grateful I am for your honesty and interest. You wonderful, smart women have changed this book and helped me "dig deeper." A special shout out to Molly, Nickie, Dara, Judy, Davie, Rachel M., Margaret, Lizanne, and Dori. Once I see the truth I can't go back. Thank you.

Julie Flanders, I know you are there, even when I can't see you. You opened my heart. David Remnitz, the best thing that ever came out of B.O.G. is our friendship; I value it so much; thanks for letting me play with the big boys.

Courtney Potts, from that first day you moved into the building, and I slipped a note under your door, you had no choice, I was going to stalk you until we became friends. You have always been my cheerleader, and you became my coach during this process. Thank you for reading all 100 (more or less) drafts with me and helping me think through them all.

There are so many friends that have supported this process and listened to all my insecurities about it, that I owe you drinks forever. To all the Jenns, Gens, and Jennys, I could not function without you. Carol, you are my sister from another mother. Suzanne, Danielle, Terri, Randi, Lauran, and Stephanie, better late in life than never. Stacy O., I love that real support comes from the most unexpected of places. To ALL the rest of my fabulous girls, way too many to name, and to one Jim Prusky, I am so grateful for your friendships, support, and just showing up for me.

To my Six-Pack (and a few more we have added along the way), I always knew I would look back on my tears and laugh, but I never thought I would look back on my laughter with tears. We are connected forever by love, memories, and now loss. Julie, you are missed every day.

Mom and Dad, I know this must have been difficult for you, and I'm so thankful for your love and encouragement, and letting me tell my version. My amazing sisters Rachel, Marissa, and Di (Deanna) . . . parts of this story belong to you too, and I could never have been brave enough to write this without your love and blessings.

Blaise, Saxton, and Grey, you are the loves of my life. The most important job I will ever have is being your mom, and I am grateful and thankful every single day for the miracles that you are. You healed me.

Whenever I have done something in my married life that most people would consider a bit (or a lot) controversial—like, ohhh, being on a reality show, or writing a memoir—most people would say to Bennett, "How are you letting her do this?" My amazing, proud husband would smile and reply, "Ahh, *let*? Have you met my wife?" Bennett, you are the truest, best, most secure person I have ever met. Thank you for knowing exactly who I am and loving me because of it. You are my best friend, I love you to death. MBT.

Live. Love. Laugh. xx Jen

About the Author

Jennifer Gilbert is the founder and chief visionary officer of Save the Date®, a New York–based special events company. She was the youngest—and the only woman—recipient of the Entrepreneur of the Year® Award in 1998, sponsored by Ernst & Young that year; she was awarded Working Woman's Entrepreneurial Award of Excellence; and her company was named among the top 500 woman-owned businesses. A frequent speaker, she has appeared at New York University, Columbia Business School, and numerous industry events. She lives in New York City with her husband and their three children. Visit her Web site at www.savethedate.com.